*Water Desalting
Planning Guide for
Water Utilities*

Water Desalting Planning Guide for Water Utilities

**Water Desalting Committee
of the American Water Works Association**

**American Water Works
Association**

WILEY

John Wiley & Sons, Inc.

Library of Congress Catagloging-in-Publication Data:

Water desalting planning guide for water utilities / Water Desalting Committee, American Water Works Association.
    p. cm.
Includes bibliographical references and index.
    ISBN 0-471-47285-9 (cloth)
    1. Saline water conversion.   I. American Water Works Association.
Water Desalting Committee.
    TD479.W38   2004
    628.1'67—dc21

                                        2003011755

Printed in the United States of America

10 9 8 7 6 5 4 3 2 1

Contents

Preface

The availability of freshwater sources to meet potable water demands is increasingly limited by population growth, resource preservation, competition, and the public's demand for higher quality drinking water. Over 40 years ago, John F. Kennedy stated:

> *If we could ever competitively, at a cheap rate, get fresh water from saltwater, . . . (this) would be in the long range interests of humanity which would really dwarf any other scientific accomplishment.*

While the United States has not yet achieved President Kennedy's vision, the worldwide humane benefits of doing so are becoming clearer with every passing decade. The use of abundant saline water sources, including seawater and brackish waters, for drinking water production continues to expand today. Limited freshwater supplies, the promulgation of more stringent drinking water standards, and technology advances in the conversion of saline water to freshwater, have made water desalting a viable, cost-effective water supply alternative for many communities in the United States and other countries.

Desalination, desalinization, and desalting are words that are interchangeably used to indicate the reduction of total dissolved solids (TDS) from a mineralized or saline water source. The saline source can be ocean water, which has a TDS content of approximately 35,000–45,000 milligrams per liter (mg/L), depending on location around the world; it can be brackish surface and groundwaters, which have TDS concentrations from about 1000 to 10,000 mg/L or more; or it can be another source of relatively high-TDS water, such as an industrial recycle or reuse stream. Desalting technologies can be used to remove salts and other dissolved constituents to produce fresh drinking water. The U.S. Environmental Protection Agency secondary drinking water standard for aesthetic quality is 500 mg/L TDS; the World Health Organization guideline is 1000 mg/L TDS maximum. It is within this range of 500–1000 mg/L TDS that the acceptability of a water source as a drinking water supply declines.

While much has been published about the application of membranes and thermal and other desalting technologies, little has been compiled on how to select the best process from among the many system procurement and implementation alternatives available. This guide is needed to provide information about the conversion of saline water sources along with cost data in an easy-to-understand format. It has been structured to benefit those who have access to saline water sources as options to conventional freshwater source development. It also provides information for those planning to use desalting technologies for other applications, such as water reclamation and reuse, and advanced water treatment for commercial and industrial purposes.

Acknowledgments

The American Water Works Association (AWWA) Water Resources Division gratefully acknowledges the work of the Water Desalting Committee in preparing this guide. Without the dedicated efforts of many individuals who volunteered many hours this resource would not have been produced. The preparation of the *Water Desalting Planning Guide* has evolved over several years. Leadership in its development has been provided by the **committee chairs,** current and former **committee members,** their current affiliations, and **outside reviewers** who contributed to this guide.

Water Desalting Committee Chairs:
Robert A. Bergman, CH2M HILL, Gainesville, FL
John T. Morris, Morris Water Resources Consultants, San Marino, CA
John E. Potts, Kimley-Horn & Associates, Inc., West Palm Beach, FL
Millard P. Robinson, Jr., Malcom Pirnie, Inc., Newport News, VA
Ian C. Watson, AEPI/Rostek, Inc., Santa Rosa, CA

Water Desalting Committee Members:
Richard M. Ahlgren, Ahlgren Associates, Waukesha, WI
John Bednarski, MWD, Los Angeles, CA
Robert S. Cushing, Malcolm Pirnie, Inc., Newport News, VA
Roy C. Fedotoff, URS Corporation, Walnut Creek, CA
Lloyd C. Fowler, Consulting Manager, Santa Barbara, CA

Edward P. Geishecker, Retired, Watertown, MA

Ernest O. Kartinen, Jr., Boyle Engineering Corporation, Bakersfield, CA

Dennis R. Kasper, Parsons Corporation, Pasadena, CA

Shannon LaRocque, Town of Jupiter, Jupiter, FL

Edward M. Lohman, Retired, Yuma, AZ

William M. McDonald, Southern California Water Company, Ontario, CA

Irving Moch, Jr., I. Moch & Associates, Inc., Wilmington, DE

Oram J. Morin, DSS Consulting, Inc., Blue Ridge, GA

Michael B. Nelson, Saratoga, CA

Kevin Price, U.S. Bureau of Reclamation, Denver, CO

James C. Reynolds, Florida Keys Aqueduct Authority, Key West, FL

William B. Suratt, Camp Dresser & McKee, Vero Beach, FL

Kenneth Thompson, CH2M-Hill, Denver, CO

Kenneth M. Trompeter, Boulder City, NV

Outside reviewers:

Kenneth Klinko, Hydranautics, Oceanside, CA

William Mills, Jr., Orange County Water District, Fountain Valley, CA

Robert Oreskovich, Dare County Water Department, Kill Devil Hills, NC

*Water Desalting
Planning Guide for
Water Utilities*

Chapter *1*

Introduction

As demand for drinking water outstrips limited fresh potable water supplies in an area, desalting brackish and other saline waters may provide an attractive alternative or supplemental water supply. Furthermore, as regulatory and public health issues drive finished water quality goals to increasingly higher standards, desalting technologies, developed for seawater and brackish sources, become options for improving water quality. In many cases, utilities are faced with a decision between desalting local saline water (such as brackish groundwater from nearby wells) or importing freshwater from a distant source. Often the use of the distant freshwater source involves very long and costly pipelines and requires obtaining ageements with multiple landowners and political entities, potential obstacles that could be difficult or impossible to overcome. On the other hand, the implementation of a desalting project also involves many critical issues, such as identifying and permitting the disposal of the concentrate from the desalting system.

This handbook is meant to serve as an overview of the topic for decision-makers, rather than as a detailed technical manual. It provides information about the major commercially available desalting processes, including process descriptions, system components, and general costs, which will be useful in feasibility evaluations of saline water supplies. The handbook also discusses issues that a utility will likely face when implementing a desalting project.

DEVELOPMENT OF DESALTING PROCESSES AND PLANT INVENTORY

Processes for desalting high-salinity water have been known for a very long time, but their use had not reached practical large-scale feasibility until the last fifty years. Significant growth in the use of desalting technologies began with technology developments in the 1950s for thermal-driven distillation systems (flash evaporators) and electrically driven membrane systems (electrodialysis) and in the mid 1960s for pressure-driven membrane systems (reverse osmosis).

According to a 1998 International Desalination Association desalting plant inventory report,[1] by the end of 1995 there were over 11,000 land-based desalting units in the world greater than 26,000 gpd (gallons per day) (100 m^3/day), having a total capacity of 5364 MGD (million gallons per day) (20,300,000 m^3/day) (Figure 1-1). Saudi Arabia was the country with the most desalting capacity (26% of world capacity); the United States was second (15% of worldwide capacity). With respect to number of units installed, the United States led with 21% of the worldwide units greater than 26,000-gpd (100-m^3/day) capacity.

Multistage flash (MSF) distillation was the desalting process with most capacity worldwide, accounting for 44% of world capacity in 1997 (Figure 1-2). Reverse osmosis (RO) ranked second with 40%, followed by electrodialysis (ED) with 6% of the worldwide capacity. As for the number of desalting plants, RO led with 63%, ED was second with 12%, and MSF was third with 10%. Sixty percent of world capacity, was used for municipal purposes.

Ranking plants by raw water supply sources, seawater plants were first in capacity (59% of the world capacity) and second in number of units greater than 26,000 gpd (100 m^3/day) capacity (30% of total). Brackish water plants ranked second in capacity (26% of world total) and first in number of units (45% of total). Of the facilities using seawater as source water, 79% of the worldwide capacity was used for municipal supplies and 15% was used for industrial users. For sources other than seawater, industrial users led with 65% of the worldwide capacity followed by municipal supplies with 17% and power plants with 14%.

In the United States, RO is the predominate desalting technology, followed by ED. According to the desalting plant inventory, at the end of 1997 there were about 2000 RO plants with a total capacity of approximately 800 MGD (3 million m^3/day) and approximately 250 ED plants with a total capacity of approximately 40 MGD (0.15 million m^3/day) in the United States.

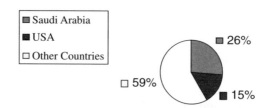

Figure 1-1 Worldwide distribution of desalting capacity.

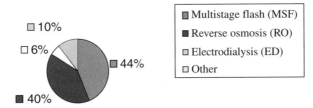

Figure 1-2 *Desalting processes used worldwide.*

HANDBOOK SYNOPSIS

The remainder of the handbook is divided into six chapters on regulations, treatment technologies, concentrate disposal, process selection, costs, and implementation, followed by a glossary of desalting process terminology and several case studies. The following is a synopsis of each chapter.

Regulations (Chapter 2)

The regulations chapter includes a brief discussion of drinking water standards, with respect to how they relate to desalting saline water sources, and applicable environmental and regulatory issues. The discharge and assimilation of concentrate from desalting processes by the environment is often the most critical regulatory issue to be faced on any project. Consequently, Chapter 4 is entirely dedicated to this topic.

Treatment Technologies (Chapter 3)

All desalting processes separate a saline feedwater stream into two streams: (1) the product, low in salts relative to the source water, and (2) the concentrate (sometimes referred to as reject or brine), high in salts relative to the source water. Numerous methods exist to separate dissolved salts, but the most significant commercially available processes can be divided into two general types: membrane and thermal.

Membrane, thermal, and other treatment technologies are discussed in Chapter 3.

With membrane technology, synthetically produced membranes are used for the separation processes. *Reverse osmosis* (Figure 1-3) is a pressure-driven process in which permeate (product water) is transported through the membrane. The membrane acts as a barrier to salts and other substances that are rejected. *Electrodialysis* and its variation, electrodialysis reversal, use an electrical potential applied across a stack of two alternating types of membranes causing dissolved salts to travel through the membranes, leaving a desalted product stream. Source water composition and the degree of salinity

Figure 1-3 *Reverse osmosis.*

removal required are important factors in selecting the most viable membrane process.

In thermal processes (Figure 1-4), influent water is evaporated, the water vapor produced is condensed to form the product water, and a concentrated brine stream is discharged to waste. The major thermal technologies are multistage flash, multiple-effect distillation, and vapor compression evaporation processes.

Concentrate Disposal (Chapter 4)

All desalting processes produce a concentrate. Regulations pertaining to concentrate discharge are complex and stringent. In some situations concentrate disposal costs and other disposal issues are so significant that they determine the overall feasibility of a desalting project. As such, concentrate disposal is a key consideration for any desalting project.

Concentrate disposal methods discussed include

- Discharge to a disposal well
- Discharge to a surface water
- Discharge to a land surface
- Discharge to a sanitary sewer and co-disposal
- Evaporation

Figure 1-4 *Multieffect distillation.*

The optimal disposal alternative for a project is site specific and disposal costs vary according to the alternative selected and the location. The options available and the approach taken vary considerably from inland to coastal environments, and depend on whether nearby receptors are fresh, brackish, or saline. Disposal methods, codisposal options, and the site-specific aspects of these issues are discussed in Chapter 4.

Process Selection and Desalting Costs (Chapters 5 and 6)

Criteria for selecting a desalting process include source water characteristics, product water goals, contaminant removal requirements, concentrate disposal issues, economics, and other factors. One of the most challenging aspects of evaluating desalting project alternatives is cost comparison. Costs of desalting technologies are constantly changing and seldom reported in a consistent manner. Concentrate disposal options and associated costs vary among desalting technologies. These and other constraints tend to make economic evaluation of desalting projects very project specific. Chapter 6 of the handbook lists key issues to consider and presents a general framework for economic evaluation as well as examples of desalting project costs.

Project Implementation (Chapter 7)

Moving from feasibility analyses and process selection to project implementation requires consideration of issues not necessarily applicable to conventional drinking water source development. Chapter 7 provides information on contracting methods of particular interest to those investigating water desalting processes.

REFERENCE

1. Wangnick, Klaus. *1998 IDA (International Desalination Association) Worldwide Desalting Plants Inventory No. 14.* Gnarrenburg, Germany: Wangnick Consulting, 1998.

Chapter 2

Regulatory Issues

The evaluation process to determine appropriate desalting technology must take into account regulatory factors that can be classified into the broad areas of drinking water quality standards, environmental concerns, and concentrate disposal. This chapter focuses on drinking water and environmental issues associated with the desalting plant itself.

DRINKING WATER STANDARDS

New municipal water treatment plants and modifications to existing treatment facilities must meet drinking water quality standards. The drinking water quality standards include *primary,* or health based, and *secondary,* which are aesthetic qualities. In addition to regulatory-enforceable primary standards, some states have elected to adopt some of the secondary standards as enforceable standards.

The U.S. Environmental Protection Agency (USEPA) established drinking water regulations for 23 contaminants under the federal Safe Drinking Water Act (SDWA) of 1974. The SDWA Amendments of 1986 required USEPA to set maximum contaminant levels (MCLs) for 83 specific constituents and to set MCLs for an additional 25 constituents every 3 years. Additional drinking water negotiations have been included in the 1996 SDWA Amendments, the Stage 1 Disinfection/Disinfection By-Product Rule and the Interim Enhanced Surface Water Treatment Rule. Additional water quality regulations continue to be developed.

Under SDWA Amendments, USEPA specifies a MCL goal (MCLG) for each contaminant it regulates and then later sets the MCL as close to the

MCLG as is technically and economically feasible. The best available technology (BAT) for achieving the MCL is also specified. Systems do not have to install the BAT to comply with the MCLs; however, systems unable to meet an MCL after installation of the BAT may receive a variance. If USEPA determines that it is not economically and technically feasible to measure the level of a regulated constituent in water, it can establish a treatment technique for the contaminant instead of an MCL.

Several national drinking water standards that are currently proposed may change before they are promulgated and others will be developed in the future. For any particular project or application, the current and proposed drinking water standards applicable to the specific jurisdictional area and project should be reviewed and complied with.

There have been many changes to drinking water standards over the years and the trend is for more stringent standards in the foreseeable future. While it is obvious that a new treatment plant must meet the drinking water standards, it is also prudent to consider future drinking water standards when planning a new treatment facility. Desalting technologies, in addition to removing salts, are capable of removing virtually all other contaminants present in a raw water source.

ENVIRONMENTAL AND SITING ISSUES

Implementing desalting facilities is similar to conventional water treatment technologies in the need to meet national, state, and local environmental regulations such as the National Environmental Policy Act, state environmental regulations, and local land use and zoning regulations. The primary environmental and siting issues related to desalting processes include

- Land use
- Noise emission
- Facility aesthetics
- Air quality and odor

The following is a discussion of each of these issues with a focus on the particular impacts created by desalting facilities. As stated earlier, many of the environmental or siting issues are common with implementation of a conventional treatment facility.

The impacts associated with all of the listed issues can change drastically, depending on the selected site for a desalting facility. A project located in an environmentally sensitive area or a sensitive neighborhood will require more effort to implement.

Land Use

Land use regulations and zoning restrictions affect a desalting water treatment facility in the same manner that they affect a conventional water treatment facility. An RO or ED treatment plant without extensive pretreatment or post-treatment facilities will occupy a small amount of space and can be designed to be compatible with a surrounding neighborhood.

Noise Emissions

RO and thermal technologies generate relatively high levels of noise emissions from the pumps, valves, air blowers, and flow restriction devices. All of these can be minimized through proper design and selection of equipment. The remaining noise emissions can be contained within the building envelope with the application of proper noise containment or absorption features. Application of full noise control to a desalting water treatment facility will increase the overall costs by 2–4%.

Facility Aesthetics

The relative compact nature of membrane desalting equipment allows it to be placed inside of a building envelope that can be designed to suit virtually any surrounding architecture. Individual process components can be separated and placed in different buildings in a "campus" style to reduce the mass of any single building. Figure 2-1 is a 5.0 MGD brackish water desalting facility on the Outer Banks of North Carolina. Intricate or elaborate architectural treatment and construction of multiple buildings will increase the overall cost of a desalting facility.

Thermal desalting plants are larger than membrane desalting plants for a given capacity. They are, therefore, more difficult to "disguise" if that should be necessary.

Air Quality and Odor

Desalting processes typically generate few if any air emissions. The combustion of fossil fuels for the generation of thermal, mechanical, or electrical energy to drive a desalting process may generate significant air emissions.

Odor emissions from a desalting facility are generally associated with the removal of gasses contained in the raw water. Hydrogen sulfide is the most common of these gasses and can lead to significant problems if not properly treated. In addition to having an extremely obnoxious odor, this gas is also toxic and significantly corrosive to electrical and electronic components. The threshold for detection by human beings is approximately 5 parts per billion, which means containment and treatment must be virtually complete, depend-

Figure 2-1   A brackish water desalting facility on the Outer Banks of North Carolina.

ing on the location of the facility. If the raw water source for a proposed desalting treatment plant contains hydrogen sulfide, the design of the facility must include components to capture and fully treat this obnoxious gas. The cost of these components will vary based on the amount of hydrogen sulfide present and could increase the cost of the desalting facility by 3–5%.

Chapter *3*

Treatment Technologies

Desalination technologies for drinking water today can be divided into two broad categories based on the underlying mechanism for removing salt molecules from water. Membrane processes use either electrical force or mechanical force (pressure) in the separation process. Thermal processes remove salt molecules by causing the solution to go through a change of phase. Virtually all thermal distillation plants operating today use boiling or evaporation to change water to a gas phase followed by condensation of that gas to a liquid containing very few salts.

This chapter discusses the more commonly used membrane and thermal desalting technologies but also includes brief descriptions of some other technologies that either are not commercially significant for public water systems (e.g., ion exchange demineralization and freezing) or are in the development process (capacitative deionization).

Membranes are used in two of the available desalting processes for drinking water treatment: electrodialysis (ED) and reverse osmosis (RO). Both of these processes use membranes to separate salts and water, but the processes use different driving forces.

RO uses pressure to separate water and salts by allowing some of the feedwater to move through a membrane, which blocks the passage of dissolved salts, thereby producing desalted water (permeate) while leaving a concentrated salt solution (concentrate) behind. Nanofiltration (NF) is a membrane process similar to low-pressure reverse osmosis and is typically used for water softening and treatment applications requiring dissolved organic hardness causing ion (i.e., calcium and magnesium) or minor salt removal.

ED employs electrical potential to move salts from the feedwater through membranes, leaving desalted water (dilute) behind as product water. Two other pressure-driven membrane processes, microfiltration (MF) and ultrafiltration (UF), do not remove dissolved salts and are not a desalting process; however, these processes are used in some cases to pretreat feedwater prior to the RO process.

Producing drinking water from saltwater sources by thermal evaporation processes (distillation) is the oldest known desalination method. Ancient Greek sailors used the concept of evaporation/condensation to produce potable water on their oceangoing sailing vessels. The distillation process is essentially the same as the natural water cycle in that saline water is heated, producing virtually pure water vapor that is condensed to form freshwater.

The typical thermal desalting process uses external energy, often in the form of steam from an electric power generating station, to heat the incoming feedwater to its boiling point. The temperature required to boil water decreases as pressure in the vessel in which water is contained decreases. By reducing the boiling point, not only will less energy be required to boil the water, but also multiple boiling steps (evaporate effects) can be utilized within the desalination plants. All three commercially available thermal desalting processes, multistage flash (MSF), multiple-effect distillation (MED), and vapor compression (VC) use these principals.

Figure 3-1 presents a simplified diagram of a desalting system. Raw water is pretreated prior to entering one or more parallel desalting process "trains."

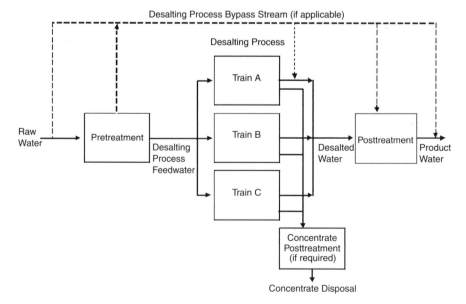

Figure 3-1 *Desalting system overall process schematic.*

Desalted product water receives posttreatment, as required for the application, and is pumped to the distribution system. The salts and other residuals separated out of the feedwater in the desalting process are continuously discharged to waste in the concentrate. If required for the specific application, the concentrate may be further treated to allow disposal.

In some applications, particularly in desalting relatively low TDS (total dissolved solids) brackish water, a portion of the feedwater may bypass the desalting process and be blended with the desalted water. This "split-treatment" design is possible if the desalted water quality is significantly better than required and blending produces a product water that still meets water quality goals. Often split treatment can reduce the required capacity of the relatively expensive desalting process part of a system and lower overall production costs.

APPLICABILITY OF DESALTING TECHNOLOGIES

Each of the desalting processes mentioned in the preceding sections has limitations and advantages. Factors such as source water TDS, plant production capacity, and association with electrical power generation will have a strong influence on which desalting process is best suited for a particular site. Table 3-1 provides a quick reference for identifying the desalting processes that are generally applicable based on various factors. This table does not include all factors and there can be exceptions to the conclusions shown there.

Table 3-1 is presented to allow the reader to identify the desalting processes that could be viable in a particular situation, or conversely, eliminate the desalting processes that that are not viable. Professionals trained in the various desalting processes should be engaged to help decide which process is most suitable.

PRETREATMENT REQUIREMENTS

Pretreatment can be defined as conditioning the raw water so that it does not damage components of the desalting process and also to reduce maintenance on the desalting equipment. A pretreatment scheme prior to all desalting processes may be required to create feedwater with the following characteristics:

- Low particulate matter or total suspended solids (TSS)
- Low scaling potential
- Low biological activity
- Low concentration of heavy metals
- Absence of elements that may oxidize to form particulate matter

TABLE 3-1 Applicablity of Commercial Desalting Processes

Factors	Desalting Processes							
	Reverse Osmosis	Electrodialysis/EDR	Multistage Flash	Multiple-Effect Distillation	Vapor Compression Evaporation	Freeze/Thaw	Ion Exchange	Capactive Deionization
Feedwater dissolved solids								
500–1000 ppm	+	+	–	–	–	–	+	+
1000–3000 ppm	+	+	–	–	–	–	–	+
3000–8000 ppm	+	+	–	–	–	+	–	o
8000–50,000 ppm	+	–	+	+	+	+	–	–
Plant production capacity								
50,000–500,000 GPD	+	+	–	–	+	+	?	o
500,000–1,000,000 GPD	+	+	–	–	+	+	?	–
1,000,000–10,000,000 GPD	+	+	–	–	o	–	?	–
10,000,000 and larger	+	+	+	+	–	–	?	–
Commercial status	+	+	+	+	+	o	+	–
Associated with electrical generation	o	o	+	+	+	–	–	–

Note. +, generally applied; o, applicable; –, not applicable.

Each of the desalting technologies differs in their sensitivity to these five basic requirements of feedwater. The extent of the pretreatment facilities, relative to both size and complexity will depend on how well the raw water meets each of the basic characteristics. Raw water that is high in suspended solids will require more extensive pretreatment facilities than raw water with low suspended solids. The following is a description of each of these characteristics, why it is needed, and the common treatment techniques.

Total Suspended Solids

Suspended solids can block the feedwater channels in a membrane desalting process or accumulate in the brine collection compartments of a thermal desalting process. Both membrane technologies, RO and ED, are significantly more sensitive to suspended solids than are the thermal technologies. Generally, thermal technologies require removal of particulate matter above ($\frac{1}{8}$ inch) while the membrane technologies require removal of all particles larger than 10 μm. All RO and ED facilities use cartridge filtration with effective removal ratings between 1 and 10 μm as a final step in pretreatment.

Both RO and ED membranes will be damaged by very small particulate matter sometimes referred to as silt. This material is usually of clay origin and will deposit on the membrane surfaces blocking the passage of water. The presence of this material is measured by the silt density index (SDI) test. Properly conditioned feedwater must have an SDI value of approximately 3 or less. Whenever possible, it is best to develop a raw water source free of suspended solids, thereby avoiding the capital and operating costs associated with pretreatment facilities.

Scale Prevention

All desalting technologies operate by separating water molecules out of the feedwater, thereby leaving behind the dissolved solids in the concentrate. The dissolved salts, or ions, can become sufficiently concentrated to join with other ions to form precipitable compounds, such as calcium carbonate, calcium sulfate, silicate, and barium sulfate. These compounds will form a scale on membrane surfaces, flow channels, or heat transfer surfaces, thereby reducing the effectiveness of the desalting process. The common pretreatment technique to prevent carbonate scale formation is addition of acid to reduce the alkalinity of the raw water.

Another technique used to prevent the formation of sulfate scale is the addition of scale inhibitor compounds. A large number of scale inhibitor compounds are available and they function by preventing the formation of any precipitable salts. This pretreatment process is common to all desalting techniques.

Biological Activity

Biological matter exists in virtually all waters and can have the same effect in a desalting process as either suspended solids or scale forming precipitants. The desalting process concentrates the biological activity and frequently provides an ideal environment for growth. Pretreatment techniques for inactivation of biological activity generally involve application of a strong oxidant, such as chlorine or ozone. This technique works well for thermal desalting technologies, but RO and ED membranes can be damaged by the presence of such strong oxidants. ED membranes can tolerate low concentrations of chlorine. RO plants using this form of pretreatment require an additional step of deactivating any remaining oxidant present in the feedwater.

Heavy Metals

The presence of heavy metals in a raw water source is generally not an issue with RO or ED desalting technologies. However, heavy metals will attack the transfer surfaces in thermal technologies and must be removed. The most common pretreatment technique is an ion trap that selectively removes heavy metals.

Particulate Formation Prevention

Certain other substances, most notably iron and manganese and hydrogen sulfide, may exist in a raw water source. If oxygen is also present in the raw water or introduced by exposing the raw water to air, or if an oxidizing chemical is injected into the water, these elements will react to form particulates. The preferred treatment approach is to prevent the introduction of oxygen into the raw water. If this is not possible, or if oxygen is already present, then sufficient oxygen and mixing energy must be introduced to completely form the particulates and then the solids must be filtered from the water.

POSTTREATMENT REQUIREMENTS

Posttreatment can be defined as conditioning the desalted water for its ultimate use, in this case drinking water. The basic requirements for drinking water are that it contain no objectionable taste or odor, that it be noncorrosive and stable, and that it be properly disinfected. Likewise, the following three basic issues should be addressed in posttreatment:

- Removal of objectionable gases
- Adjustment of alkalinity and pH
- Disinfection

The following is a description of the treatment processes associated with each of these issues and how they achieve the basic requirements of posttreatment.

Gas Removal

Generally, membrane desalting technologies allow the gases entrained or dissolved in the feedwater to pass through to the product water. Typically these gases are hydrogen sulfide, carbon dioxide, and oxygen. Hydrogen sulfide requires removal to avoid taste and odor in the finished water. Carbon dioxide will create an unstable product water. Elevated levels of oxygen will increase corrosivity. The most common process for removing these gases is forced draft degassification.

Alkalinity and pH Adjustment

All of the desalting technologies produce desalted water with very low alkalinity since (the bicarbonate ions are not carried through to the desalted water). This leaves an unstable and generally low-pH water, which would be corrosive and usually objectionably soft. Posttreatment techniques include pH adjustment by the addition of caustic soda and alkalinity increase through the addition of lime.

A technique for increasing the alkalinity of the product water is split treatment or bypassing the desalting process with some of the feedwater to blend with the desalted water. This may be possible since the desalting technologies yield very low-TDS desalted water. The proportions of bypass water and desalted water are adjusted so that the (blended) product water meets the water quality goals. The bypassed feedwater generally may be included in the degasification process. The use of split treatment requires careful examination of the contaminants that might be present in the feedwater and their impact on quality of the product water.

Disinfection

All drinking water must be disinfected to assure that no biological contamination is present in the product water and to maintain an adequate level of disinfectant in the distribution system.

MEMBRANE TECHNOLOGY

Reverse Osmosis

The RO process uses a semipermeable membrane to separate dissolved ions and water molecules. A feedwater pump elevates feedwater pressure based on the water's dissolved solids concentration, desired recovery, and specific membrane performance. Typical RO feed pressures are shown in Table 3-2.

TABLE 3-2 Typical RO Feed Pressures

Application/RO Membrane Type	Pressure Range
Seawater desalting–seawater membranes	800–1200 psi (5520–8280 kPa)
Brackish water desalting	
Standard (medium-pressure) membranes	300–600 psi (2070–4140 kPa)
Low-pressure membranes	150–300 psi (1030–2070 kPa)
Ultra-low-pressure membranes	100–150 psi (690–1030 kPa)

RO membranes are assembled in pressure vessels containing 1–7 spiral-wound membrane elements. The vessels are placed in series and in parallel to produce the desired amount of permeate (desalted) water. In many cases multiple vessels are arranged in parallel, forming a "stage." Commonly, the concentrate from the first stage is fed to additional stages of modules, depending on the design system recovery.

An alternative RO configuration is 2-pass, where the permeate from the first pass is fed to a second RO pass for further desalting. This is seldom done with brackish water because a single RO pass usually yields a low enough TDS. It is more often practiced with seawater desalting.

An assembly of vessels is termed a train. Figure 3-2 shows a 2-stage train, containing 37 vessels in the first stage and 14 vessels in the second stage

Figure 3-2 *Two-stage train located in Jupiter, Florida. Reprinted courtesy of Kimley-Horn & Associates, Inc.*

with 6 spiral-wound elements in each vessel. This train treats brackish water and produces 1.5 MGD product water.

Control   Instrumentation on an RO system typically includes sensors for pH, pressure, electrical conductivity, and flow for operational control. The concentrate control valve regulates flow to establish overall recovery of the train. Instrumentation can vary from manual to completely automated systems. As a minimum, an RO system should include instrumentation to shut itself down when any operating parameter deviates significantly from normal to prevent damage to the membranes. These shutdown parameters include feedwater pH, feed and concentrate pressure, permeate electrical conductivity, and permeate flow.

The degree of automation and instrumentation beyond these minimums is usually a function of the complexity of the system, number of operating process trains, and the requirements of the facility owner or operator.

Reliability   RO is a reliable process for water supply when properly designed and operated. Generally, it is recommended that municipal water treatment plants have at least two RO trains to allow for down time for membrane cleaning and other maintenance without shutting down the entire desalting system. Available finished water storage and water demand characteristics should drive decisions about the number of trains and total size of the facility. RO trains are usually operated at a relatively constant flow rate. Matching water demand with plant production requires multiple trains and/or adequate storage. Figure 3-3 shows simplified schematics of typical reverse osmosis process configurations.

Theory   Most living organisms use natural osmosis to transfer fluids across cell or tissue linings. When a semipermeable membrane separates two fluids that have different concentrations of dissolved solids, water moves from the lower concentration fluid through the membrane to the side with higher concentration to dilute it. The amount of pressure required to stop the flow of water molecules from the fluid of lower concentration to the fluid of higher concentration is the *osmotic pressure*. The movement of water molecules across the semipermeable membrane can be reversed by applying pressure to the fluid having a higher concentration of dissolved solids. As the difference in the dissolved solids concentration between the two fluids increases, the osmotic pressure increases. A useful rule of thumb is that osmotic pressure increases 1 pound per square inch (psi) (6.9 kPa) for each 100 mg/L difference in dissolved solids concentration of the fluids on each side of the membrane.

The principle of RO is based on applying pressure greater than the osmotic pressure to a high TDS fluid. This causes water to pass through the semipermeable membrane from the high TDS side to the lower TDS side while the membrane rejects the salts.

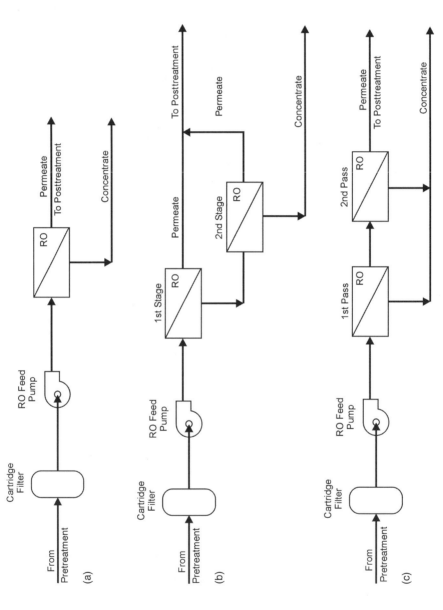

Figure 3-3 Typical RO process schematics: (a) single-stage, single-pass; (b) two-stage, single-pass; (c) two-pass.

Membranes Pressure drives water through the membrane. As driving pressure increases, the rate of water molecules crossing the membrane increases. The rate at which water molecules move across the membrane is called *flux*, generally expressed as gallons per day per square foot of active membrane surface, gal/day/ft² or gfd (cubic meters per second per square meter, $m^3/s/m^2$ or m/s). Water flux is the measure of the membrane productivity.

A combination of net driving pressure and flux produces one of the more important measures of membrane performance. The amount of pressure required to force a given amount of water across the membrane is called specific flux, expressed as gfd/psi. Specific flux is a significant factor in determining the cost to produce drinking water using a particular membrane.

Some dissolved solids pass through the membrane while a majority do not. The percentage of dissolved solids that do not pass through the membrane is termed rejection. Membranes are available that reject 99.5% of the dissolved solids. The method by which dissolved solids pass through a membrane and why membranes reject different species of ions (calcium, sulfate, sodium, etc.) differently is complex. The molecular weight, mobility, and electrical charge of the ions play a role. However, the passage of dissolved solids through the membrane does not change as the driving pressure varies, as is the case for water molecules. The significance of this is that as flux decreases, TDS of the permeate increases. This is because the mass (pounds) of dissolved salts that pass through the membrane is contained in a smaller volume of water.

A flat sheet RO membrane rejecting barrier is generally less than 0.2 μm thick and could not withstand a significant pressure on only one side without a support surface. To overcome this, the membrane is cast onto a porous backing sheet that provides structural support to withstand a relatively high pressure differential. These flat sheets are then rolled in a spiral configuration as shown in Figure 3-4.

Recovery The RO process can extract only a portion of the water molecules available in a source water. The percentage of feedwater extracted as permeate is called the *recovery*. Recoveries typically can range from 60 to 85% of the feedwater supply for brackish water, and 30 to 50% for seawater.

Two factors limit recovery in an RO system: formation of solid compounds (scale and precipitate) and driving pressure. As water molecules pass through the membrane leaving behind dissolved solids, the solids rejected by the membrane concentrate on the feedwater (concentrate) side of the membrane. As dissolved solids concentrations increase, scaling can occur as described in the Pretreatment section. It is possible to increase recovery by using chemical additives in the feedwater, usually acid or scale inhibitors.

As water passes through the membrane leaving behind dissolved solids, the solids concentration in the feedwater (concentrate) increases. This, in turn, increases osmotic pressure. There are limitations to the driving pressure that

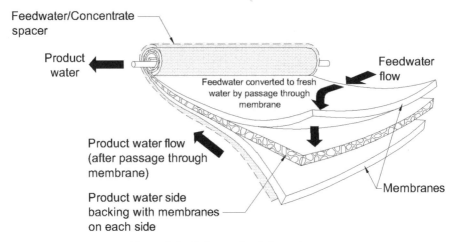

Feedwater/Concentrate spacer

Product water

Feedwater converted to fresh water by passage through membrane

Feedwater flow

Product water flow (after passage through membrane)

Product water side backing with membranes on each side

Membranes

Figure 3-4   *Cutaway of spiral membrane.*

can be applied to a membrane, backing sheet, pressure vessel, and other components.

Advantages and Disadvantages   Through proper membrane selection, many contaminants can be removed from a source water. In addition to rejecting dissolved substances, RO membranes also provide an effective barrier to viruses, bacteria, and other pathogens (giardia and cryptosporidium).

Small-capacity RO plants are compact, enabling them to be placed in a small area, and they are often installed near the source water or the end users. RO plants are commonly designed with multiple parallel trains with individual train capacities up to about 3 MGD or more. The RO trains are usually housed in a building.

Operations and Maintenance   RO does not require a "settling in" period during start-up. The individual train can be started and stopped periodically but care must be taken to prevent damage to the RO system.

An RO plant can be automated to reduce operating costs. Controls can start and stop the system based on tank level, shut down the plant or a train if operating parameters deviate from normal, and record operating data. However, there are limits to automation. For instance, it is difficult to have automatic controls load chemicals, clean analyzer probes, or clean filters.

Over time, the membrane feedwater channels become fouled and chemical cleaning is required. Eventually, the membrane elements require replacement when cleaning does not yield acceptable membrane performance.

Electrodialysis

Electrodialysis is an electrochemical separation process in which ions are transferred through ion selective membranes by means of a dc voltage (Figure

3-5). Electrodialysis reversal (EDR) is a modification to ED where the polarity of dc power is reversed several times per hour. The most common application for ED/EDR is the reduction of total dissolved solids in brackish waters to meet drinking water standards. TDS removal from feedwater by ED/EDR systems can range from less than 15% to about 90%, depending on the system configuration and applied electrical current.

Theory   When electrodes are placed in a solution of dissolved solids and dc power is applied, ions will migrate toward the electrode with the opposite charge of the ion, as shown in Figure 3-6. Cations (Na^+) are attracted to the cathode (negative electrode) and anions (Cl^-) are attracted to the anode (positive electrode).

To desalt water, the movement of ions (TDS) is controlled by the addition of membranes that form watertight compartments, as shown in Figure 3-7. Two kinds of membranes are used in electrodialysis. The anion transfer membrane (A) allows only the passage of negatively charged anions (chloride, sulfate, bicarbonate, nitrate, for example). The cation transfer membrane (C) allows only the passage of positively charged cations (calcium, magnesium, sodium, potassium, for example). The membranes are electrically conductive and are essentially impermeable to water under pressure.

Figure 3-5   *Electrodialysis plant.*

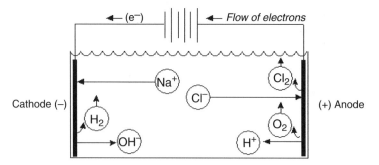

Figure 3-6   *Sodium chloride solution under dc potential.*

Figure 3-7 shows what happens when dc potential is applied across the electrodes. The figure shows six compartments separated by the ion-selective membranes. The applied dc voltage exerts a force on ions promoting migration toward the electrode of opposite charge. The membranes limit how far ions can travel. Compartments 1 and 6 contain metal electrodes, which provide the needed electromotive force. Compartments 2, 4, and 6 are depleted ions. Anions, such as Cl^- ions, pass through the anion membranes (A) into compartments 3 and 5, while cations such as Na^+ ions move through the cation membranes (C) into compartments 1, 3, and 5. Compartments 1, 3,

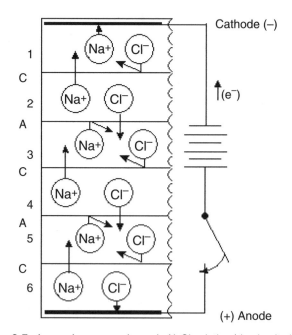

Figure 3-7   *Ion exchange membrane in NaCl solution (dc circuit closed).*

and 5 contain a concentration of cations and anions that move through the anion and cation membrane array.

Process Description

Continuous Electrodialysis   In practice an electrodialysis system is composed of a series of stacks. Each stack has an inlet and an outlet for each space between the membranes. These inlet and outlet channels are created by manifolds in the compartments. The desalted water (dilute) makes a single pass through the stack and exits as desalted water. The concentrate stream is partially recycled to minimize waste, as shown in Figure 3-8.

At the system outlet, there is a blowdown to concentrate disposal. The amount of concentrate makeup controls recovery. A stack is made up of cell pairs. A cell pair consists of a cation transfer membrane, a dilute flow spacer, an anion transfer membrane, and a concentrate flow spacer. A typical electrodialysis membrane stack may have from 300 to 600 cell pairs.

Electrodialysis Reversal   EDR is an improvement over ED in that scaling and chemical addition are greatly reduced while maintaining continuous production of dilute by reversing the electrical polarity of the applied dc power 2–4 times each hour. When the polarity is reversed, the dilute stream and concentrate stream compartments are reversed. This alternating exposure of membrane surfaces to the dilute and concentrate streams provides a self-cleaning method that enables desalting of waters that would normally scale or foul. This self-cleaning capability allows many EDR plants to operate without continuous chemical feeds.

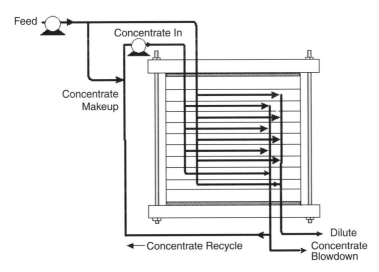

Figure 3-8 *Continuous electrodialysis.*

Configuration of Facilities   Typical EDR systems employ more than one stage. Each stage typically removes up to 50% of the feedwater salt content based on salt composition and the water temperature. Most systems are designed to have 2–4 stages, as shown in Figure 3-9, to achieve the desired salt removal. The concept of staging leads to great flexibility in system design with standard components. To increase the amount of salt removal, more stages are added. To produce more dilute, lines (trains) of stacks are operated in parallel.

Operation and Maintenance   ED/EDR plant operations may be highly automated and can offer long-term reliable operation. ED/EDR systems can be cleaned in place to restore system performance. Mineral scales are removed by circulating a weak acid solution through the membrane stacks. Organic fouling is removed by circulating a caustic solution through the membrane stacks. Severe fouling or scale can be remedied by disassembling the membrane stacks and manually cleaning the membranes with little or no loss in membrane life. The membrane stacks can be sanitized by circulating a chlorine solution through the system.

Advantages and Disadvantages   As a general rule, salinities greater than about 2000 mg/L are not economically desalted by ED/EDR since the energy usage is proportional to the amount of salt removed. ED/EDR does not remove non-ionized substances, such as many particles, organics, and microorganisms (viruses, giardia, and cryptosporidium, for example). Therefore, EDR can treat water with a relatively high level of particles and turbidity as compared to RO. Also, water recovery is not limited by silica in the feed, because EDR does not remove silica.

THERMAL PROCESSES

The introduction of flash evaporators in the 1950s resulted in the development of several large-scale distillation processes that were suitable for commer-

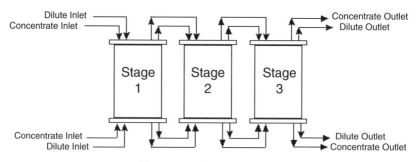

Figure 3-9   *Stages in series.*

cial operation. The majority of thermal desalination plants in use today are used for seawater desalination. In these processes, multiple evaporation and condensation stages (or effects) are used to generate desalted water or distillate. Thermal processes can produce distillate water with a TDS of below 10 mg/L.

The evaporator performance ratio, also referred to as the gain output ratio, is probably one of the most common terms in use to compare various evaporative desalination plant designs. The performance ratio provides a measure of the pounds of distillate produced per pound of motive steam applied to the process. For example, the single-effect evaporation process, which is used in small industrial applications, returns slightly less than one pound of distillate for each pound of steam. The most common evaporators consist of the large-scale, multistage flash units (MSF), which have been widely installed in the Middle East. These MSF plants typically deliver between 2 and 8 pounds of distillate for each pound of steam. Higher economy evaporators, including horizontal multieffect distillation (MED) plants exhibit performance ratios that range upward to approximately 16 pounds of product delivered for each pound of steam. Recent innovations in MED technology include a vertical arrangement of the evaporative effects (vertical tube evaporator, VTE), and this process, termed the VTE-MED, may achieve performance ratios of up to 24 pounds of desalted water per pound of steam.

Water needs two important conditions to boil: (1) it must reach the proper temperature relative to its ambient pressure, and (2) there must be sufficient additional energy to ensure vaporization of the liquid. When water is heated to its boiling point and the heat is turned off, vaporization of steam will not occur because additional energy (the heat of vaporization) must be applied to the water to boil it and produce steam vapors. Boiling and vapor evolution can be maintained by adding more heat or by reducing the ambient pressure above the water. If the pressure is reduced, as occurs in an MED process when the feedwater passes from one evaporative effect to the next, the water is at a temperature above its boiling point and vapor evolution occurs as "extra" heat is added to the process. As the heat of vaporization is applied, the temperature of the water falls to the new boiling point because the steam produced takes heat out of the remaining liquid water.

Aside from multiple boiling effects, another important factor in thermal processes is scale control. Although most naturally occurring substances dissolve more readily in warmer water, some substances dissolve more readily in cooler water. Unfortunately, some of these substances like carbonates and sulfates are found in seawater. One of the most important substances related to scale formation is gypsum ($CaSO_4$), which begins to leave solution (precipitate into solid form) when water approaches about 203°F (95°C). This material forms a hard scale that can coat evaporator tubes and vessels. Scale formation inhibits heat transfer across the evaporator tubes and reduces the efficiency of the distillation process. Scale formation also causes mechanical problems in the unit by plugging openings and, once formed, scale deposits are difficult to remove. One way to avoid scale formation is to keep the water temperature and water chemistry within prescribed limits.

Various forms of thermal desalination have been successfully applied in locations around the world. The thermal process that accounts for the most desalting capacity is the multistage flash (MSF) distillation process; the second most widely used thermal process in terms of installed capacity is multiple-effect distillation. A third distillation process in use is vapor compression evaporation (VCE). These processes are described further on.

Large thermal desalination plants are often combined with power generation stations. Since most large thermal desalination plants require steam as the driving force, thermal processes typically make economic sense when incorporated into a dual facility in which both electrical energy and desalted seawater are produced concurrently or in utility or industrial situations where significant quantities of exhaust or previously used heat are available.

Multistage Flash

In the MSF process, water is heated in a vessel called the brine heater to about 240°F (115°C). In the brine heater, steam passes over a bank of tubes containing the feedwater. The heated feedwater then flows into another vessel called a "stage" in which the pressure is lowered such that the water entering the stage will immediately boil or flash into steam. Generally, only a small percentage of the heated water that enters the stage will flash to steam. The exact conversion quantities depend on the pressure maintained in the stage, since boiling continues only until the feedwater cools to the boiling point at that pressure. Using this process, feedwater can pass from one stage to another and be boiled repeatedly without adding more heat because the pressure in each stage is less than the pressure in the preceding stage. Today, a typical MSF plant contains 4–40 evaporative stages.

Steam vapors generated by the flashing process are converted to freshwater by condensing the vapors on heat exchanger tubes that run through each stage. The heat exchanger tubes are cooled by incoming feedwater on its way to the brine heater. This action, in turn, heats the feedwater so that the amount of thermal energy needed in the brine heater to raise the temperature of the feedwater is reduced. Figure 3-10 presents a simplified schematic of the MSF process.

In addition to once-through MSF plants, "recycle" configurations are used so that part of the concentrate (brine) is mixed with the incoming feedwater and returned to the concentrate. Also, separate heat recovery and concentrate sections are incorporated in the recycle configuration. The recycle configuration is used to reduce pretreatment costs and improve the process performance ratio by reducing heat loss.

MSF Theory   MSF evaporators operate on the concept of releasing vapor from a boiling liquid by introducing it as a superheated liquid into a chamber maintained at a pressure low enough to allow boiling to continue without introduction of additional heat energy.

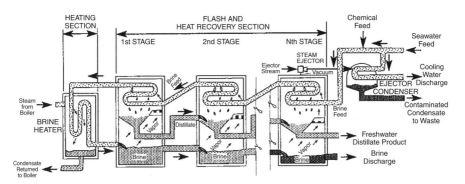

Figure 3-10 MSF process schematic.

Advantages and Disadvantages   Through the multiple staging of the MSF cycle, performance ratios in the range 10–15 pounds of product water per pound of applied motive steam can be achieved. Typically, however, most MSF plants operate with performance ratios of less than 10. Operation of the plant at conditions different than design will result in changes in the pressure drop or flow rate ineach stage. Large changes cause unstable water levels in each stage and can lead to reduced performance ratios or distillate contamination. Plants with a significant number of flashing stages, say 30 or more, are more sensitive to changes from design conditions than those with less than 30.

Operations and Maintenance   Thermal desalination systems are a complex accumulation of vacuum systems, steam systems, pumps, and compressors. They are amenable to automatic control, and, in fact, it is recommended to simplify operation.

Important parameters that must be monitored and controlled for correct operation include (among others) recovery (feed, product, and concentrate flows), temperatures, and pressures within the effects. If these parameters are not properly controlled, it is possible to damage the heat transfer surfaces either with scale (requiring significant downtime and effort to remove the scale) or corrosion (requiring replacement of the tubes).

Regular maintenance is required to keep a thermal system operating. There may be a product pump for each effect, as well as concentrate and feed pumps. In many cases these pumps operate at high temperature or on a corrosive liquid. Steam ejectors, steam generators, vacuum pumps, and valves also require continued maintenance. Even with proper operation, tubing leaks occur, requiring regular (typically annual) shutdown and repair of the heat transfer systems. The long experience with thermal desalination plants in the Middle East and the Caribbean demonstrates their ability to provide long-term service.

Multiple-Effect Distillation

In the past decade, renewed interest in the MED process has produced a number of new high-efficiency MED designs. Most of these new units have been built around the concept of operating at low temperatures (up to 170°F).

The MED process, like the MSF process, takes place in a series of vessels (also referred to as "effects") and uses the principle of reducing the ambient pressure in the various effects as the temperature decreases. This concept allows multiple boiling of feedwater and condensation of product distillate without supplying additional outside heat after the first effect. In an MED plant, feedwater enters the first effect and its temperature is raised to the first boiling point after initial preheating in feedwater tubes. Feedwater is then either sprayed or otherwise distributed onto the outside surface of tubes in a thin film to promote rapid boiling and evaporation. The tubes are heated on the inside by motive steam from a powerplant or by steam from a boiler. The condensed motive steam is recycled to the power plant or boiler for reuse. Only a portion of feedwater applied to the tubes in the first effect is evaporated into the pure water vapor. The remaining feedwater is fed to the second effect where it is again applied to the outside surfaces of a tube bundle, which are heated by vapor created in the first effect. This vapor is condensed in the second effect to form distillate. As distillate is formed, heat is given up to evaporate a portion of the feedwater in the second effect. This process continues for several more evaporative effects. Typically, 8–16 effects are used in a large MED plant. Due to the horizontal arrangement of the evaporator tubes, feedwater remaining in each effect must usually be pumped to the next effect to apply it to the next tube bundle. A schematic of the MED process is shown in Figure 3-11.

Theory   Steam generated from boiling saline water in each of the effects becomes the motive or driving stream for subsequent effects. The feedwater to the first effect is heated with steam from a boiler while steam generated in the last effect is sent on to a final condenser as in a single-effect evaporator. Operating in this manner requires decreasing the pressure in steps from the

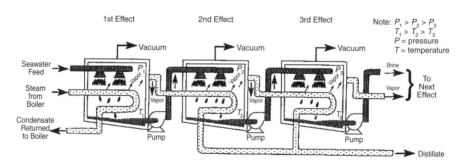

Figure 3-11   MED process schematic.

first effect through the condenser. This pressure differential is the driving force that draws steam from any effect to the next effect in series. Liquid flow is likewise cascaded from effect to effect.

The performance ratio of the single-effect evaporator is limited to slightly under one pound of distillate from each pound of motive steam. The practical limitation of MED performance ratio is approximately 10–24. This level of process efficiency is typically accomplished with 15 or more evaporator effects in series. In practice, the performance ratio of most commercially available MED evaporators in seawater desalination applications is limited to less than 15 pounds of distillate per pound of steam due to economic considerations.

Advantages and Disadvantages   Major advantages of the MED process, as compared to MSF, include higher performance ratio and lower operating temperatures. Using the same temperature difference per effect and designing for a performance ratio of 12 can be accomplished in an MED plant with about 15 effects at a maximum brine temperature of 170°F, compared to the MSF process maximum operating temperature of 240°F.

Another advantage offered by MED is that it can be configured in a vertical design (VTE-MED), which dramatically reduces the plant footprint and minimizes land costs. The heat exchange rates in a VTE-MED plant can be much higher due to the falling thin-film evaporative process. Increased heat exchange rates result in lower heat transfer tube bundle costs, a significant factor in overall plant costs. Recent innovations in a vertically arranged MED unit promise to achieve performance ratios in excess of 20.

Vapor Compression Evaporation

The vapor compression evaporation distillation process is generally used for small-scale (less than 5.0 MGD) desalting units. The heat for evaporating the water comes from the heat resulting from compressing the vapor, rather than the direct exchange of heat from steam. Like the MSF and MED processes, the VCE process is also designed to take advantage of the principle of reducing the boiling point temperature by reducing the process pressure. The primary method used to produce heat to evaporate incoming feedwater is a mechanical compressor. The mechanical compressor is usually electrically driven, which eliminates the need to obtain or generate steam.

Figure 3-12 is a simplified diagram showing a mechanical vapor compression evaporator. The compressor creates a vacuum in the vessel and compresses vapor removed from the vessel. As vapor leaves the vessel, it condenses on the inside of a tube bundle releasing heat. Feedwater is sprayed on the outside of the heated tube bundle, where it boils and partially evaporates, producing more vapor.

In some applications, a steam jet is used in lieu of a mechanical compressor. In this process, also referred to as a thermo vapor compressor, a venturi

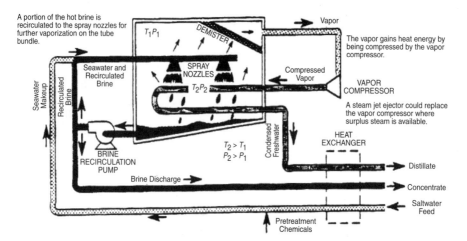

Figure 3-12   *Mechanical VCE process schematic.*

orifice at the steam jet creates and extracts water vapor from the main vessel creating a lower ambient pressure in the main vessel. Extracted water vapor is compressed by the steam jet. This mixture is condensed on tube walls to provide the thermal energy (heat released by condensation of the vapor) to evaporate feedwater applied on the other side of the tube walls.

Theory   The VCE cycle uses steam produced by the evaporation of water and increases it to the higher energy level by compressing the vapor by either mechanical means or thermodynamic energy. In the compression cycle, an extremely high fraction of the energy can be used to achieve a high performance ratio.

Mechanical Vapor Compression   The most common mechanical vapor compression (MVC) cycle uses single-stage centrifugal compressors designed to bring the steam through a compression ratio of approximately 1.5–2.0. Mechanical compression cycles generally operate most practically at pressure levels at, slightly above, or slightly below the atmospheric boiling point. Thus, the highest MVC performance ratios can be achieved with mechanical equipment that delivers only a small pressure increase (temperature increase) to the steam. The disadvantage of this situation is that the net driving force for heat exchange within the equipment is small. This, is turn, requires that evaporator tube bundles be constructed with large heat transfer surface area. In general terms, MVC processes require more complex mechanical cycles and greater heat transfer surface area than do equally as efficient MED or MSF plants.

Thermo Vapor Compression Thermo vapor compression (TVC) is most commonly used to enhance the performance ratio of multiple-effect VCE systems. In certain applications, TVC can double the performance ratio of multiple-effect systems. Once again, system losses must be considered and in this case additional losses or inefficiencies due to the steam ejector compressor operation must also be considered. The most practical advantage of TVC is that it boosts the comparatively low economies of multiple-effect VC systems to a higher or intermediate economy range without adding evaporator effects. A disadvantage of TVC systems is that they require moderately high-pressure steam sources to drive the ejectors. The advantage, relative to MVC is, that no moving parts (i.e., mechanical compressors) are required, therefore, maintenance requirements of the TVC process are generally less than for the MVC system.

Advantages and Disadvantages When source water is extremely high in dissolved solids, distillation by VC can be utilized, because this process can accommodate high dissolved solids concentrations.

Operations and Maintenance In addition to the primary energy source (steam, electrical, or mechanical) required to operate an evaporative desalination system, auxiliary energy is usually required. Electrical energy is required for the recirculation, feed, and blowdown pumping system. High-pressure steam (approximately 150 psi or higher) for evaporators operating under vacuum conditions is usually required as motive energy for steam ejector-driven vacuum systems. Mechanical vacuum pumps are normally not used except for the smallest packaged evaporators.

OTHER DESALTING PROCESSES

The preceding pages addressed the more commonly used water desalting processes for public drinking water supplies (reverse osmosis, electrodialysis, and distillation). This section briefly describes some of the other desalting processes: ion exchange demineralization, freezing, and capacitive deionization (CDI). Ion exchange is commonly used for low TDS demineralization for industrial applications. Freezing and CDI are not yet commercially significant.

Ion Exchange Demineralization

Ion exchange demineralization with anion and cation exchange resins can be used to remove essentially all dissolved ions from water. Positively charged ions are exchanged for hydrogen ions and negatively charged ions are exchanged for hydroxide ions. The hydrogen and hydroxide ions form water molecules. Cation and anion exchange resins can be placed in separate vessels (dual-resin bed system) or in a single vessel (mixed resin bed). The amounts of hydrogen and hydroxide ions on the cation resin and anion resin, respectively, are finite. Only so much feedwater can be treated before the hydrogen and hydroxide ions are depleted and must be replaced in order to desalt more feedwater. This is done by passing acid through the cation resin to replace the hydrogen ions, driving off the cations absorbed from the feedwater during the desalting cycle. Caustic soda is used to replace the anions absorbed by the anion resin. The cost of acid and caustic regenerants preclude ion exchange from being economical for high-TDS removal applications, such as brackish or seawater desalting.

Freezing

Several freezing processes have been demonstrated for desalting water. Freeze desalting is based on the premise that ice crystals, formed when saltwater is frozen, are essentially free of salt. It takes only 13.5% as much energy, 144 Btu/pound, to convert water into ice as it does to convert water into vapor, 1070 Btu/pound. Several processes using these principles have been developed, including direct freezing, indirect freezing, and absorption (Figure 3-13).

Direct Freezing—Vacuum Freezing Vapor Compression Direct freezing methods involve reducing the feedwater temperature to near its freezing point, creating a vacuum and condensing the salt-free vapor that results. Vacuum

Figure 3-13 *Freeze-thaw processing equipment.*

freezing vapor compression (VFVC) is one variation on this method. It requires maintaining a vacuum as well as transport and compression of large volumes of water vapor.

Indirect Freezing The indirect freezing process (IFP) was developed to overcome the problem of VFVC by using a refrigerant with a much higher vapor pressure than water. The refrigerant must, of course, be immiscible with water so that water can be separated from the refrigerant. The IFP is quite similar to the VFVC process except that it operates at a higher pressure and uses a refrigerant.

Absorption Water can be desalted using a hygroscopic material, which is a substance that absorbs and retains water such as lithium chloride. Absorption processes are based on water vapor being absorbed by the hygroscopic material, which is kept cold during the absorption step by heat exchange with melting ice. Heat is applied to the hygroscopic material, driving off water as a vapor. The hygroscopic substance is then ready for reuse.

Capacitative Deionization

Capacitative deionization (CDI) was developed at the Lawrence Livermore National Laboratory in California. The CDI process is based on the use of a substance called *carbon aerogel*, a highly porous and extremely light solid sometimes called frozen smoke. It has a large surface area per unit volume of carbon aerogel. A desalting device using CDI would, in appearance, be

Figure 3-14 *Carbon aerogel test apparatus.*

much like an electrodialysis stack (Figure 3-14). Carbon aerogel is fastened to both sides of a metal plate. A number of these plates, separated by spacers and gaskets, are fastened together in a stack. Electrodes anodes and cathodes are placed on either end of the stack.

In operation, feedwater flows through the stack between the metal plates holding the carbon aerogel. Dissolved solids or ions in the feedwater are drawn into the carbon aerogel by the electrical attraction between the electrodes and ions. Water exiting the other side of the stack has fewer ions (dissolved solids) in it than feedwater. Trapped ions are released into a relatively small stream of rinse water. CDI is not presently commercially used to desalt water.

Chapter *4*

Concentrate Disposal

INTRODUCTION

When evaluating any of the desalting technologies, it is critical during the planning stages to determine conclusively the method for disposing of concentrate. Regulations pertaining to concentrate discharge are complex, stringent, and require careful attention. It is essential to determine the full cost of concentrate disposal when planning a desalting facility. Proper planning and thorough knowledge of concentrate discharge issues can help to minimize disposal costs.

Water desalting plants generate concentrate whenever they produce desalted water. For large plants, the concentrate must be discharged as it is generated. For smaller plants, it is possible to detain the concentrate and discharge it in a slug fashion, but this changes only the rate of concentrate discharge, not the volume.

Disposal may include treating concentrate to make it compatible with the ultimate discharge location or may require transporting it a significant distance. When these requirements are added, the cost and operating costs increase greatly. In general, the best disposal solutions involve identification of a suitable saline sink as the ultimate receiving body.

CONCENTRATE DISPOSAL METHODS

This section provides a general description of regulation and treatment requirements involved in the five basic methods of concentrate disposal as well as some innovative methods. Disposal is site specific and can be intricate or simple. In addition to local geography and environment, federal, state, and local regulations play a large role in choosing a method for concentrate disposal. The five basic methods of concentrate disposal are

- Injection well to underground aquifer
- Discharge to a surface water
- Discharge to a sanitary sewer
- Land application
- Evaporation

Discharge to an Injection Well

This method of concentrate disposal involves injecting concentrate into an acceptable underground aquifer using a disposal well. An acceptable aquifer is one that contains water of a similar character to the concentrate. Regulations generally require that the concentrate not degrade the waters of this aquifer or that the aquifer has no beneficial use and therefore can withstand degradation. The obvious key requirement to implementing this method of concentrate disposal is that there be an aquifer with these characteristics in close proximity to the desalting facility. There are several advantages of this disposal method:

- The injection well can generally be located in the vicinity of the treatment facility, thereby minimizing pipeline costs.
- Treatment of the concentrate prior to injection is generally not required.
- The concentrate is "out of site and out of mind," which means that it is most likely no longer of public concern.

The following are several major disadvantages of this disposal method:

- Acceptable aquifers may be several thousand feet deep. This combined with required construction techniques makes each well cost $2,000,000 to $4,000,000.
- An injection well is considered a mechanical device and must be taken out of service periodically for testing. This generally requires construction of at least two injection wells to provide redundancy.
- This disposal method eliminates the use of concentrate as a future resource.

There are a number of desalting water treatment plants operating in South Florida using this method of concentrate disposal. The typical well extends approximately 3000 ft (900 m) into a porous geological formation containing what is essentially seawater. Geologic formations isolate this saline aquifer from drinking water aquifers, which lie above.

Discharge to Surface Water

This method of disposal includes a wide variety of specific applications, since surface water bodies have an infinite variety of characteristics. The basic variations are size, flow rate, salt content, turbidity, usage, and applicable regulations.

The essential element in this method of concentrate disposal is the availability of a receiving water body that will not be adversely affected by the concentrate. Discharge of concentrate to a surface water body has the following advantages:

- The discharge facilities are generally low in cost.
- The components of a surface water discharge are reliable.

Two disadvantages of this method of concentrate disposal are that

- The permitting process is relatively complicated.
- Generally, the concentrate must be treated or diluted prior to discharge to surface water.

The first step in evaluating disposal to a surface water is to determine the nature and quality of the concentrate. The next step is to identify surface water bodies available with suitable water quality. Minor incompatibilities can frequently be resolved by treating the concentrate to remove or modify the constituent causing the incompatibility. It is necessary to investigate flow characteristics of the potential receiving water bodies to determine if adequate mixing and dilution of the concentrate will occur.

Discharge to a Land Surface

This method of concentrate disposal is generally limited to percolation ponds and use of concentrate as an irrigation water. Advantages of this disposal method include the ability to use concentrate as a resource, the potential for revenue if used as an irrigation resource, and relatively low overall costs. Disadvantages include the need for large areas of land and the responsibility to produce a consistent quality of concentrate. Evaluating concentrate for use as an irrigation water must include determining the effects that constituents in the concentrate will have on the vegetation irrigated and the underlying groundwater. This can be complex, since vegetation differs in tolerance of dissolved solids, and the long-term effects of some concentrate constituents may require lengthy testing.

Disposal to a percolation pond requires relatively large land areas, depending on the permeability of the soil and, most importantly, the compatibility of concentrate with underlying groundwater. This method should be evaluated in the same way as discharge to surface water because of the need for compatibility between concentrate and underlying groundwater.

Discharge to a Sanitary Sewer

This method of disposal is limited to discharging concentrate into the collection system of a wastewater treatment facility. The advantages of this method include relatively simple facilities at the water treatment plant, and generally no pretreatment of the concentrate. Disadvantages include relatively high capital costs in the form of connection fees at the wastewater treatment plant, using up hydraulic capacity, and continuing high operating costs in the form of wastewater treatment fees. Evaluating this method for concentrate disposal involves examining concentrate volume compared to wastewater treatment capacity, and the compatibility of the concentrate quality with the wastewater treatment process and treated wastewater disposal method. If the wastewater treatment agency practices or is contemplating recycling the treated wastewater for beneficial use, it will probably prohibit the additional salt load from the concentrate discharge.

Evaporation

Evaporation is effectively a zero-discharge option since the concentrate stream is reduced to solids through evaporation. The advantages of this disposal method include zero discharge, low operation costs, and generally no concentrate treatment requirements. Disadvantages include the need for relatively large land areas and periodic disposal of accumulated solids. This disposal method is limited to areas of arid climate, generally the western United States. Evaluating evaporation requires determining the area required to evaporate all concentrate under the most adverse climate conditions.

INNOVATIVE APPROACHES

Concentrate disposal is not limited to the five commonly used methods discussed previously. Sites may have available unique and innovative methods of concentrate disposal that cannot be characterized in a manual of this type. The key to developing unique and innovative disposal methods is to consider the concentrate a resource and make an open-minded evaluation of potential users or discharge locations in the vicinity of the desalting plant. This requires a thorough understanding of the concentrate's character and potential discharge locations or the potential value of selected constituents to a nearby industry.

Discharge for Reuse by Another Facility   Concentrate can be discharged to a facility to be used as cooling or process water. When this disposal method is used, the discharging facility usually does not have to modify the concentrate. Instead, the facility using the concentrate may modify it specifically for their use through dilution or chemical treatment.

Codisposal Codisposal refers to discharging concentrate into an existing waste stream. This could be a domestic treated wastewater outfall, thermal waste stream, or industrial waste discharge. Since concentrate may be classified by regulatory agencies as an industrial waste, the codisposal waste stream may also be classified as industrial. This method of concentrate disposal has the potential benefit of using existing discharge facilities and existing permits. If concentrate causes significant changes in the character, quality, or quantity of the codisposal stream, then existing facilities and permits may require modification.

Applicable Regulations

Discharges to any groundwaters are regulated by the Underground Injection Control (UIC) provisions of the SDWA. USEPA generally administers the UIC program, but some states have been delegated the right to administer these regulations.

The UIC regulations establish five classes of wells, two of which are applicable to concentrate disposal. Concentrate is currently classified as an industrial waste, which most frequently requires the use of a Class I or a Class IV well. Class IV wells are banned because they are a direct threat to underground sources of drinking water.

These regulations prohibit discharges into an aquifer unless it can be demonstrated that the injected fluid will not degrade the receiving aquifer. UIC regulations also include specific requirements for construction of an injection well aimed at preventing leakage of the injected fluid. In general, these construction regulations require fully cementing all casing, construction of the well with multiple stages of casing, and installation of a monitoring well network to detect leaks.

The Clean Water Act regulates discharges to surface waters through the National Pollutant Discharge Elimination System. Regulations generally allow for mixing zones in a receiving water body to dilute contaminants and toxicity. Mixing zones allow discharge of concentrate containing elevated levels of contaminants and acute or chronic toxicity. It is necessary to have a thorough knowledge of regulations when evaluating this issue. Although the issue of toxicity seems costly and time-consuming to resolve, the cost benefit of discharge to a surface water body can be significant.

TREATMENT REQUIREMENTS

Concentrates are treated to meet a regulatory requirement at the point of discharge. As previously discussed, an almost infinite variety of locations and conditions make it impossible to identify specific treatment processes and equipment. Concentrate does not usually contain significant levels of suspended solids. However, thermal process concentrates may contain more solids than membrane process concentrates, since extensive membrane processes require essentially solids-free feedwater.

Generally, no treatment of concentrate is required prior to its disposal into an injection well. Concentrate discharged to surface water may require some level of dissolved oxygen to minimize any adverse impact on aquatic life. In-line air sparging or aeration of the concentrate stream are the most common methods of adding dissolved oxygen.

Many concentrates also contain hydrogen sulfide, toxic to aquatic life and a significant odor nuisance when the concentrate is exposed to air. Hydrogen sulfide can be removed through aeration or reduced by oxidation. Aeration imposes an energy requirement, while oxidation imposes a chemical requirement.

Some concentrate constituents may be toxic. The safe level of concentration for some toxic substances is so low that they cannot be detected in source water until it is concentrated. It may be necessary to perform a pilot test to obtain a concentrate sample for analysis.

Treatment requirements must be evaluated with disposal options. Objectionable constituents of concentrate are specific to the discharge method and location. Practical and cost-effective treatment techniques to remove these constituents must be developed as a function of evaluating each discharge alternative. The cost of operations and maintenance associated with concentrate treatment facilities must be clearly identified. A proper evaluation balances the capital and operating costs of concentrate disposal facilities with the other concentrate disposal alternatives.

Chapter *5*

Process Selection

Factors that need to be considered in selecting the desalting process most appropriate for a given situation include

- Desalter utilization
- Treated water quality goals and removal requirements
- Source water quality
- Pretreatment considerations
- Contaminant removal
- Recovery
- Product water posttreatment
- Process pilot testing
- Blended treatment
- Waste and concentrate disposal
- Energy usage and availability
- Site development and environs
- Costs (see Chapter 6)
- Environmental and regulatory constraints (discussed in Chapter 2)

Depending on conditions at a specific site, there may be other process selection factors than those listed. However, the basis for process selection in most cases is the *least cost per gallon of acceptable product water.*

Factors to consider for process selection are site specific; some more so than others. For example, the cost of desalting equipment for a specific source water, treated water quality goals, and a given production rate is generally not affected by site conditions. But the cost of energy and waste disposal can vary greatly depending on the conditions prevalent in a specific case. And, of course, site development and environmental and regulatory costs are also site specific.

DESALTER UTILIZATION

The production capacity of a desalting plant is critical not only in the cost of the facility, but also in the process selection. For example, membrane plants are common for small capacity desalting plants but not distillation processes. Desalting plants are typically constructed in parallel modules (sometimes called process "trains"), which easily allow phased capacity addition to meet growing demands. Membrane processes (e.g., RO and ED) as well as ion exchange and vapor compression are better suited for phased construction for most plants, because they are more cost-effective in smaller capacity modules than MSF or MED processes.

How the desalter water will be used affects the design of the project. If, for example, the desalter is to serve a "base load" and will be utilized at a constant, continuous delivery rate, then the design and operating protocol may be different from that of a "peaking" facility, which will operate intermittently and, perhaps, at varying production rates. If the desalter is to provide peaking capacity, it should be noted that startup of a thermal desalting plant takes time, where as a membrane plant yields desalted water immediately upon startup.

TREATED WATER QUALITY GOALS

Treated water quality goals influence the entire treatment process selection (pretreatment, desalting, and posttreatment) for any specific case. Assuming that the objective of the desalting project is to supply water for municipal use, applicable drinking water standards and the requirements of the public health authorities having jurisdiction must be met.

More stringent water quality goals than those required by regulatory agencies may be set in some instances. For example, a water purveyor may wish to provide water of the same quality from a desalting plant as provided from the purveyor's other water sources even though the resulting desalted water quality may be better (and more expensive) than health authorities or prevailing law may require. Or the objective may be to blend desalted water with water from other sources. If the water to be blended has high salinity, the desalted water needs to have low salinity so that blended water meets water quality goals. If the water to be blended is of adequate quality, the desalted water may have higher salinity with the resulting blend still meeting water quality goals. Product water quality, desalted only or a blend of desalted and other water, must be carefully considered. Generally, the more stringent the product water quality, the higher the treatment cost.

SOURCE WATER QUALITY

A primary step in selecting a water treatment process is to locate the most reliable, highest quality source water. Ideally, the source water will be protected from deterioration through a watershed/groundwater protection program or natural features. The quantity and quality of source water directly affects water treatment process selection.

When *fresh* source water quantity is limited or distantly located, or quality is too poor for more conventional treatment without supplemental treatment processes, saline water sources are often considered. The same issues as to source protection and reliability will also then apply to the saline source.

Source Water Analyses

To determine source water quality, representative samples should be taken often enough and for a long enough time period to include all known variables and to allow projection of estimated changes in water quality over time. For example, source water quality analyses should include the constituents shown in Table 5-1. If the desalting facility incorporates a bypass stream, as is typical for brackish water sources, microbiological constituents also need to be assessed.

Hydrogeologic Investigations

These studies are used to define the characteristics of potential groundwater sources. They are particularly important for brackish aquifers whose quality may change as water is extracted over time. The source water quality characteristics at initial plant start-up and throughout the life of the facility must be estimated for design. The investigation should also verify that an adequate quantity of source water is available and consumptive water use permits are attainable.

PRETREATMENT CONSIDERATIONS

Pretreatment may be designed to remove dissolved substances to protect process equipment materials, to remove suspended solids and colloids from the source water if necessary, and to add chemicals such as scale inhibitor and/or acid.

The type of desalting process selected impacts the choice of pretreatment. For example, if a distillation process is used for desalting, source water does not need to be as free of suspended and colloidal solids as it does for the

TABLE 5-1   Recommended Minimum Source Water Quality Data

Membrane Desalting Applications		
Temperature (field)	Iron[a]	Hydrogen sulfide
pH (field)	Manganese[a]	Silica
Total dissolved solids	Sulfate	Radionuclides
Conductivity	Chloride	Color
Hardness	Barium	Bromide
Alkalinity	Strontium	Turbidity
Calcium	Fluoride	Silt density index[b]
		Hydrogen sulfide
Magnesium	Phosphate	Chlorine/strong oxidants
Sodium	Ammonium	Total organic carbon
Potassium	Nitrate and nitrite	Total bacterial plate count[c]

Thermal Desalting Applications		
Temperature (field)	Iron	Hydrogen sulfide
	Copper	
pH (field)	Manganese	Silica
Total dissolved solids	Sulfate	Radionuclides
Conductivity	Chloride	Color
Hardness	Barium	Bromide
Alkalinity	Strontium	Turbidity
Calcium	Fluoride	
Magnesium	Phosphate	Chlorine/strong oxidants
Sodium	Ammonium	Total organic carbon
Potassium	Nitrate and nitrite	Total bacterial plate count[c]

[a] Total and dissolved.
[b] For pressure-driven membrane processes only.
[c] Heterotrophic plate counts.

membrane processes, but it may need to have heavy metals removed. For a surface water source (for example, an open seawater intake), single-stage sand filtration may be adequate for a distillation desalting process, although even that may not be required depending on the level of suspended solids in the source water. If, however, reverse osmosis is to be used for desalting surface water, then two-stage media filtration, membrane filtration, or a coagulation-sedimentation-filtration process, is most likely needed to achieve low suspended and colloidal solids. If the source water is groundwater from wells that essentially is free of suspended solids, cartridge filtration may be all that is needed for pretreatment solids removal. Electrodialysis is typically more tolerant of feedwater suspended and colloidal solids than reverse osmosis.

The degree of chemical pretreatment needed depends on source water quality, desired recovery, and the desalting process selected. Acid is often added to reduce the alkalinity of the feedwater to prevent the formation of calcium carbonate scale and allow increased recovery. An antiscalant (scale inhibitor chemical) is also often used to prevent precipitation and scale formation in a desalting process and allowing recovery to be maximized. Table 5-2 presents

TABLE 5-2   Typical Pretreatment Requirements

Process	Operating Temperature	Source Type[a]	Pretreatment Requirements[b]
Multistage flash distillation	190°F (88°C)	S or G	Antiscalant
Multistage flash or multiple-effect distillation	235°F (113°C)	S or G	Acid or antiscalant, deaeration, ion trap[c]
Multiple-effect distillation	170°F (77°C)	S or G	Antiscalant, ion trap[c]
Vapor compression distillation	Ambient	S or G	Antiscalant, ion trap[c]
Vapor compression distillation	190°F (88°C)	S or G	Antiscalant, ion trap[c]
Reverse osmosis or electrodialysis	Ambient	G	Antiscalant and/or acid, cartridge filtration[d]
Reverse osmosis or electrodialysis	Ambient	S	Direct filtration (1 or 2 stage) or membrane filtration or coagulation-sedimentation-filtration, antiscalant and/or acid, cartridge filtration[d]

[a] S, surface water; G, groundwater.
[b] In addition, disinfection is required for drinking water systems (usually chlorination).
[c] Ion traps are needed to remove heavy metals when aluminum heat exchange tubes are used.
[d] Additional pretreatment may be required to remove specific contaminants such as iron, manganese, or silica.

generalized pretreatment requirements for the major desalting processes. Pretreatment processes generate waste streams, which need to be disposed of. The methodology and costs of the disposal also need to be identified during the planning stage.

CONTAMINANT REMOVAL

Desalting process selection is dependent on the target contaminant removal requirements, which are known after the source water is characterized and product water quality goals are defined. Desalting processes exhibit different contaminant removal capabilities. A generalized summary of the removal capabilities for regulated contaminant classes potentially in drinking water by various desalting technologies is presented in Table 5-3.

Inorganics

All desalting processes effectively remove dissolved ions or total dissolved solids (TDS). Thermal processes typically separate nearly all dissolved minerals, leaving product water TDS less than 10 mg/L in many cases. Reverse osmosis systems typically remove 90–99% of the TDS, depending on

TABLE 5-3   Generalized Regulated Contaminant Removals by Desalting Technologies

Regulated Contaminant	Distillation (%)	Reverse Osmosis (%)	Electrodialysis (%)
Inorganics	>99.9	90–99[a]	50–90[b]
Pesticides and synthetic organic compounds	50–90[c]	90–99	<5
Volatile organic compounds	50–90[c]	5–50[c,d]	<5
Chlorinated organics	50–90[c]	5–50[d]	<5
Microbiological	>99	>99	<5
Radiological	>99	90–99	50–90

[a] Seawater-type RO membranes can reject >99% of salts such as NaC1.
[b] ED does not remove uncharged compounds, such as inorganic silica.
[c] However, nearly complete removal is possible with *common* pre- and / or posttreatment processes.
[d] Removal depends on type of organic and membrane selected.

membrane type and operating parameters. Electrodialysis systems remove high levels of ionized substances, but not unionized dissolved solids such as silica.

Organics

The thermal desalting processes may volatilize organics in the feedwater and carry them into the distillate if the boiling point of the organic is close to that of water. Occasionally, a distillation plant may have a granular activated carbon filter as a posttreatment process for removal of volatile organics that may be present. Most organics are rejected by reverse osmosis membranes. Electrodialysis is not a barrier process and organics in the feedwater stream generally pass through the ED process and remain in the desalted water.

Color

Color in water can be caused by the presence of organic or inorganic materials. Organic color-causing contaminants may evaporate in a thermal process. RO, because it is a barrier process, removes color-causing contaminants. ED and EDR are generally not effective in removing color-causing organic contaminants from water.

Microorganisms

Viruses, bacteria and other pathogens may be "inactivated" (killed) by the high-temperatures in a distillation process, especially high-temperature processes over 200°F (93°C). However, some health authorities may not accept distillation as meeting the Surface Water Treatment Rule. Reverse osmosis, on the other hand, since it is considered a barrier process, is usually given high credit for virus, bacteria, giardia, and cryptosporidium removals. Slight imperfections in the membrane barrier and potential O-ring leaks could allow individual particles, such as microbes, to enter the permeate flow stream, so posttreatment disinfection is needed for potable water systems. Research is currently in progress to develop accurate and reliable on-line integrity test methods for spiral-wound membrane systems. Electrodialysis does not provide a barrier and does not remove microorganisms.

RECOVERY

Product water recovery, the ratio of product to feedwater flow, is an important consideration in selecting and designing a desalting process. There are several reasons to maximize recovery:

- Increasing recovery reduces the quantity of the source water, which is especially important in water short areas.
- Higher recovery requires less pretreatment facility capacity and less pumping for a given product flow rate.
- Increased recovery rates result in less concentrate, but of a higher concentration, which may affect disposal costs.

Seawater Recoveries

For seawater desalting, recoveries range from about 10% for a multistage flash distillation plant without recycle to 30% with recycle. Recent research and development with the multieffect distillation process has yielded recoveries on the order of 66%. Mechanical vapor compression distillation plants recover about 45% and reverse osmosis (RO) plants recover about 45–50% for seawater desalting applications with 1400-psi RO membranes, seawater desalting plants can attain 60% recovery. Electrodialysis is not commonly used for seawater desalting because of the high required electric power consumption.

Brackish Water Recoveries

For brackish water desalting, recoveries depend on the type of desalting process and water chemistry, especially the presence (or absence) of potential scale-forming minerals. Recoveries of 65–85% are not uncommon for brackish water reverse osmosis plants. Electrodialysis plants, particularly electrodialysis reversal plants, generally can be designed for a slightly higher recovery than reverse osmosis for a given feedwater. Thermal processes are not typically used for brackish water applications because of their relatively high costs.

PRODUCT WATER POSTTREATMENT

Pure water is referred to as the universal solvent. That is, water without sufficient dissolved minerals (TDS) of the right mix tends to dissolve materials that it comes in contact with. Removing dissolved minerals from the source water may decrease the stability (increase the corrosiveness) of desalted water. It is usually necessary to treat the desalted water to reduce its corrosivity. Posttreatment corrosion control may be as simple as adding caustic soda to raise the pH or adding a corrosion inhibitor such as zinc orthophosphate. It is very common to install a product water degassifier prior to chemical addition, to remove carbon dioxide (thereby raising the pH and minimizing alkaline chemical addition).

The specific posttreatment needed for a particular case needs to be selected carefully. Potential blending with other waters in the distribution system must be considered to prevent unwanted reactions at the mixing interface. Besides corrosion control, all desalting plants for public water supplies require product water disinfection. In some cases, other processes are used in posttreatment. For example, sometimes granular activated carbon filters are included for organic removal. If gases (e.g., hydrogen sulfide) or volatile organics compounds are of concern, degassifiers or air strippers may be included in posttreatment. For seawater desalting, it is common to add alkalinity and hardness to the desalted water because the desalted water contains mostly sodium chloride. The water is corrosive and has little, if any, taste. Adding hardness and alkalinity addresses both of these problems.

PILOT TESTING

Pilot testing of a process may be conducted to verify and optimize design criteria, operational parameters, and expected full-scale system costs; select acceptable products; determine the effects of site-specific conditions and pretreatment requirements; produce samples of product and concentrate for analytical testing; and for other reasons. The decision to pilot test should consider the costs of process risks and uncertainties versus the cost of pilot testing. Sometimes the need for testing is lessened if there is similar process experience at another site treating the same source water.

SPLIT TREATMENT

In cases where the desalted water quality is significantly better than needed to meet finished water quality goals, it may be acceptable and desirable to use split treatment. If possible for a given application, raw or pretreated water bypass/blending is advantageous in that it typically lowers the required capacity of the desalting equipment, reduces the waste concentrate flow rate, minimizes posttreatment corrosion control, and results in lower capital and operating costs.

WASTE DISPOSAL

A desalting plant produces two kinds of waste, liquid and solid. The quantity and quality of the liquid waste stream depends on source water quality, pretreatment and desalting processes, product water quality goals, and recovery achieved in the desalting process. For instance, where pretreatment filtration is required, the quantity of solids from the filtration process depends on the suspended solids content of the source water and the type of filtration process employed. Typically, backwash water from a filtration process uses 2–10% of the filtered water. For a more extensive pretreatment facility, such as one incorporating chemical coagulation, sedimentation, and filtration, sludge from the clarifiers and backwash water from the filters must be disposed of. The desalting process itself, generates a waste concentrate flow stream. Chapters 2 and 4 discussed waste disposal permitting and concentrate disposal methods, respectively.

ENERGY USAGE

All desalting processes require energy to separate the dissolved solids (salts) from the water. Most thermal processes use steam to supply at least a substantial portion of their energy needs. Mechanical vapor compression desalters have been designed to operate with only electric motor or internal combustion engine power. Reverse osmosis and electrodialysis membrane desalting processes also normally use electric power. Steam or internal combustion engine-driven pumps can be used for membrane process desalting plants. For electrodialysis, electric power is always required even if non-electric motor-driven pumps are used to move water through the system because electricity is needed as the desalting driving force.

Energy requirements for desalting plants are typically thought of as being large. Compared to conventional water filtration plants, this may be true. However, energy requirements vary considerably among desalting processes. Table 5-4 compares the energy requirements of the several desalination process discussed. To compare the energy requirements, thermal (steam) and electrical power demands were converted to equivalent barrels of oil. For purposes of preparing the table, it was assumed that a barrel of oil contains 6,000,000 Btu, the overall efficiency of an oil-fired power plant is 40%, and boiler efficiency (to make steam) is 80%. Using these assumptions, steam and electric power requirements for the desalting processes were converted to barrels of oil as shown in the table.

TABLE 5-4   Comparison of Approximate Energy Requirements for Desalting Processes

Process	Approximate Energy Requirements		Equivalent Barrels of Oil/MG
	Steam (Btu/MG)	Electric Power (kWh/MG)	
Brackish water			
RO	0	5,500	8
EDR	0	7,300	10
Seawater			
MSF	830,000,000	12,000	190
MED	710,000,000	10,000	162
VC	0	30,000	43
RO	0	16,000	23

Note. RO, reverse osmosis; EDR, electrodialysis reversal; MSF, multistage flush; MED, multieffect distillation; VC, vapor compression.

The figures in Table 5-4 indicate why reverse osmosis rather than thermal seawater desalting processes have been used around the world except in those cases where very inexpensive energy is available, such as Saudi Arabia where large supplies of natural gas are available at very little cost and desalting plants are often adjacent to or part of a power plant process so that steam is readily available. To give perspective to the energy requirements shown in the table, consider the energy requirements to import freshwater. For example, the California State Water Project imports water from Northern California to as far south as San Diego. It takes approximately 10,000 kWh/MG (equivalent to 14 barrels of oil) to deliver raw water to Southern California and treat it. This does not include the energy needed to pressurize the treated water for delivery to customers.

Another interesting comparison is that the average single-family home in California consumes the equivalent of about 25 barrels of oil per year. This figure does not include energy used outside the home for such things as transportation. Assuming that a single-family home uses 100,000 gallons of water per year, the total energy requirement for a home in California would be equivalent to about 27 barrels of oil per year if imported water is used or about 28 barrels per year if desalted seawater were used. From these figures, more energy is required to desalt seawater than to import and treat freshwater. However, the overall increase (about 4%) in energy consumption, if desalted seawater rather than imported freshwater were used, would be relatively small as compared to the total energy consumption of a single-family home.

Advances in desalting process technology and design of facilities can be expected to result in reductions of energy requirements in the future. A great deal of attention is being given to energy requirements by the entire desalting industry. The operating pressures for RO membranes continue to decline, energy recovery in RO processes continue to advance, and innovation in the thermal desalting technologies continues to reduce capital as well as operating costs. Operating pressures for brackish water reverse osmosis membranes have been reduced by 50% or greater over the last 20 years. The VTE-MED concept discussed in Chapter 3 offers to make a substantial improvement in the performance ratio of thermal desalting. The frequent refinements being made in all of the desalting technologies underscore the need to involve an experienced professional in the early stages of planning a desalting water treatment facility.

Since desalting processes require energy and because thermal desalting plants rely on heat energy, these plants are often built in conjunction with power plants. Steam to drive the thermal process is extracted from the turbine driving the electrical generator. For gas turbine or diesel driven power plants, waste heat boilers are placed in the turbine or diesel exhaust to recover heat.

SITE DEVELOPMENT

Site development is a broad term intended to cover the cost of developing the site for a desalting plant. Development costs include bringing energy to the site, land clearing/grading, land area requirements, subsurface conditions, access, distance to the water source, distance to the point(s) where the desalted water is needed, and other such considerations. A major consideration is the availability of energy to operate the desalting plant. If a desalting plant requiring steam is to be built, either steam-producing equipment is needed as part of the plant design or the desalting plant needs to be built adjacent, or very near, to an existing steam source such as a power plant. If electricity only is needed to operate the desalting plant, the cost of conveying power to the desalting plant site needs to be considered. Topography is important. If a site has considerable relief, site grading, foundation and structural costs are greater than if a level site is selected.

Land area requirements are very important in water treatment plant planning and feasibility analysis, particularly in highly developed areas where land values are high. With respect to size, membrane desalting plants requiring no or minimal pretreatment filtration are generally smaller than conventional filtration plants of the same capacity. If, however, the feedwater contains solids so that pretreatment filtration is required ahead of the desalting process, the area required for the plant could increase significantly and conceivably be as large, or larger, than that required for a conventional filtration plant.

Small or medium plants that use distillation to desalt water typically require more land area than membrane desalting plants. Table 5-5 presents generalized land area requirement comparison for the primary desalting technologies for a 10-MGD (38,000 m^3/day) product capacity plant.

Geotechnical considerations are also important in selecting a site. For areas with poor foundation conditions, the weight of the desalting process equipment and the ancillary equipment needs to be considered, along with the weight of any enclosing structures. Access to the site for both construction

TABLE 5-5 Typical Land Area Requirements for an Assumed 10-MGD (37,850-m³/day) Desalting Plant

Technology	Source Water (Application)	Approximate Area
Distillation (MSF)	Seawater	2.5 acres (10,000 m²)
Distillation (VTE MED)	Seawater	1.5 acres (6,000 m²)
Distillation (VC)	Seawater	2.0 acres (8,000 m²)
Reverse osmosis	Seawater	2.0 acres (8,000 m²)
	Brackish	1.0 acres (4,000 m²)
Electrodialysis	Brackish	1.0 acres (4,000 m²)

and operation of the completed plant must be considered in selecting a site. Traffic normal to the operation of a complete desalting plant must also account for chemical deliveries and solid waste removal.

The cost of bringing source water to the desalting plant and conveying concentrate off site may be significant, since the source water and disposal site may be some distance from the desalting plant site. Similarly, the cost of conveying treated water from the treatment plant to consumers must also be included in the planning/feasibility stage. Substantial source and treated water conveyance, storage, and pumping facilities may be needed as part of the complete desalting project.

For a water system supplied by numerous brackish water wells scattered throughout the water service area, the costs of a central plant versus plants at the wellheads should be considered. While a single, large, central plant might be less expensive than several small wellhead plants, the cost of pipelines to convey source water from the wells to the treatment plant and the treated water to consumers may be more expensive.

Aesthetics may be an important consideration. It may be necessary to construct a desalting plant so that it does not look like an industrial facility. Horizontal process equipment may need to be concealed within buildings and vertical process equipment may need to be disguised through creative architecture.

Chapter *6*

Desalting Costs

An important topic that is of interest to anyone considering a desalting project is the cost. An extremely important concept when discussing costs for a desalting project is the need to consider *all* of the costs. That is, what should be considered is not just the cost of the desalting plant, but the cost of the *desalting project*. As is discussed in the following pages, there are more often than not facilities other than just the desalting plant must be constructed to make desalted water available to customers.

The cost of a desalting project depends on many factors, most of which are peculiar to a specific site. This section describes many items that need to be considered when estimating the cost of a desalting project. However, for any specific project the cost items described here may not include all of the cost items that should be considered.

DEFINITION OF CAPITAL COST AND OPERATION AND MAINTENANCE COST

Two basic types of costs are associated with a desalting project:

- Capital cost
- Operation and maintenance (O&M) cost

Capital Cost

Capital cost includes two cost components:

- Construction—The cost of equipment, buildings, pipelines, and other physical facilities that make up the project

55

- Incidentals—The cost of engineering, environmental, legal, administrative, and financial services needed to plan, design, permit, and construct the project

Construction cost is usually by far the majority of the capital cost. The magnitude of the construction cost depends on a great many variables, such as the following:

- Source water quality
- Product water quality goals
- Size (capacity) of the desalting project
- Means of conveying source water to desalter
- Means of conveying the product water to customers
- Means of disposing of residuals (concentrate, brine, and solids)
- Site development issues, such as,
    Availability and cost of land
    Geotechnical conditions
    Site development
    Land use
    Architectural constraints
    Environmental/permitting requirements
    Other community concerns
    Availability (and cost of) power
    Access

Operation and Maintenance Costs

O&M cost includes the costs of actually operating and maintaining the desalter and producing desalted water. O&M costs include such cost components as power, labor, chemicals, membrane replacement, concentrate disposal, and miscellaneous (such as repairs and replacement). In some cases, it may also be necessary to purchase the water to be desalted.

O&M cost may be further broken down into two categories—fixed and variable costs. Fixed O&M costs are those costs that are independent of the amount of water desalted. Variable O&M costs are those costs that are dependent on (vary with) the amount of water that is actually desalted. For example, labor costs (including wages/salaries and fringe benefits) are usually fixed costs, whereas paper and chemicals are variable costs. The energy costs for desalting plants can vary greatly depending on the salinity of the source water and type of plant, the unit power cost (electricity, oil, or other fuel), the use or energy-recovery devices, and other factors. As an example, typical seawater desalting plant O&M costs for a 5-MGD (19,000-m³/day) are about $1.50 to $2.50 per 1000 gallons of product water. Typical O&M costs for a similar-sized brackish water desalting plant are approximately $0.50 to $1.00 per 1000 gallons.

CONSTRUCTION COSTS

Figure 6-1 presents typical construction costs for brackish and seawater de-salting plants in U.S. dollars per gallon per day installed plant capacity. The seawater costs are for *single-purpose* plants, where water production is the only product from the facility. Colocating with a power generation facility can save significantly on both the capital and operating costs for a desalting facility. All desalting processes can benefit from the availability of the cooling water for source and the outfall for concentrate disposal. Thermal processes also benefit from steam availability. For single-purpose seawater applications, distillation and RO facility costs are shown. ED is normally not economically competitive for TDS applications greater than about 3500 mg/L TDS (sea-water is about 35,000 mg/L TDS). Typical seawater desalting plant construc-tion cost for a 5-MGD (19,000-m³/day) capacity plant is about $3.00 to $6.00 per gallon per day installed capacity.

For brackish water applications, only RO and ED processes are shown; thermal processes are not cost effective for lower TDS drinking water de-salting applications. The brackish water costs shown in Figure 6-1 assume a

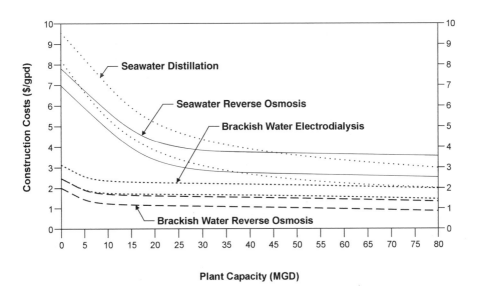

Figure 6-1 *Construction costs for desalting plants.*

source water TDS of about 3000 mg/L. Typical brackish water desalting plant construction cost for a 5-MGD (19,000-m³/day) plant is about $1.00 to $2.00 per gallon per day installed capacity.

The costs indicated by Figure 6-1 are "on-site construction" costs only. They do not include incidental (engineering, legal, financial, etc.) costs or contingencies. Neither do they include "off-site" costs for such improvements as conveying feedwater to the site, transporting product water to customers, bringing power to the site, and so on. The cost curves also assume no unusual geotechnical, architectural, environmental, etc., conditions that could significantly increase the construction cost.

ESTIMATING CAPITAL COSTS

Following are brief discussions of the factors that affect capital costs.

Source Water Quality

Two major aspects of source water quality impact the cost of a desalting project:

- The presence of inorganic (silt, for example) and organic (algae, etc.) solids
- The presence of dissolved substances, including TDS and other dissolved inorganic and organic contaminants

The temperature of the source water is important also. As the water temperature decreases, the cost of desalting increases due to increased operating pressure for membrane desalting and increased energy (heat) for thermal plants, for example.

Product Water Quality Goals

In general, the more stringent the product water quality goals, the higher the cost of desalted water will be. The minimum product water quality goal must be to meet the applicable drinking water standards (assuming that the desalter will supply drinking water). However, there may be situations in which meeting the drinking water quality standards is not good enough. For example, if the existing water supply is of lower TDS or hardness than required by the drinking water quality standards, the product water quality goal may be to match the existing water supply quality. For a seawater reverse osmosis desalting plant, this may require a "partial second pass" to provide a product water with sufficiently low TDS to match the existing water supply. This second pass will increase the cost of water from the desalter.

Another reason that a TDS lower than required by drinking water regulations may be needed is because there may be restrictions on the TDS of the treated effluent from the local wastewater treatment plant. For the wastewater plant to meet its discharge requirements, it may be necessary to desalt the product water to a lower TDS than is necessary to meet just the drinking water quality requirements. This, too, will increase the cost of desalted water. Setting the product water quality goals for a desalter should be carefully thought through so that better water quality than necessary is not sought, thereby unnecessarily increasing costs.

Conveying Source Water to the Desalter

Delivering the water to be desalted to the desalter can be a significant part of the construction cost of a project. For instance, in one brackish groundwater-desalting project, the construction cost of the wells and pipelines needed to pump and convey the brackish groundwater to the desalters was estimated to be about 30% of the total estimated construction cost of the project. For a seawater-desalting project, the cost of developing the feedwater supply was about 10% of the estimated project construction cost. Including the cost of developing the water supply for a desalter can be a significant portion of the total capital cost and must not be overlooked when planning a project.

Delivering Product Water to Customers

As with developing the source water supply for a desalter, delivering the product water to customers can be a significant cost. For the groundwater desalting project used as an example above, the cost of delivering the groundwater desalter's product water to customers represented about 17% of the estimated project construction cost. For the seawater-desalting plant example, the cost of facilities to deliver the product water to customers was estimated to be about 20% of the total construction cost. As with the cost of developing the water supply to the desalter, the cost of delivering the product water to customer can be a substantial percentage of the total capital cost.

Disposing of Concentrate/Brine

All desalting processes produce a wastewater (concentrate or brine). Disposing of the concentrate can be costly. Permitting and environmental requirements can be difficult (expensive) to meet depending on the circumstances peculiar to a particular situation. (See Chapter 4.)

For the groundwater-desalting project discussed above, the cost of purchasing capacity in an existing brine disposal line was equivalent to 30% of the estimated construction cost of the desalter. For the seawater example, an existing treated wastewater outfall pipeline could be used. The cost of connecting to this pipeline to dispose of the concentrate was equivalent to only about 2% of the project construction. These two examples point out the broad range of differences that may be encountered in disposing of concentrate or brine. The initial cost for disposal can be a substantial portion of the total project construction cost.

Site Development

Some site development issues are straightforward. For example, it can readily be determined whether the parcel of land upon which the desalter is to be built is large enough. The availability and cost of property can be ascertained. Whether constructing a desalting plant is compatible with surrounding land uses is also easily determined. However, it is more difficult to determine other aspects of construction costs associated with a particular property. Consider, for instance, geotechnical suitability. The seawater desalter used as an example above would have been sited adjacent to San Francisco Bay in an industrial area. The property was owned by the project's owner, was undeveloped, and was large enough for the purpose. However, a pile foundation would have been required. The estimated cost of the pile foundation was about 10% of the project's construction cost. This same situation was experienced at a brackish groundwater-desalting project. Instead of being able to use shallow foundations (spread footings), some type of deep foundation was needed. The least expensive deep foundation was driven piles. Driven piles added about 5% to the construction cost as compared to shallow foundations.

Architectural constraints can also impose cost increases as compared to an industrial building design. Seawater desalters by their very nature are located at or very near the seashore. The need for desalted seawater indicates that the surrounding area is probably developed, perhaps very densely. Locating an industrial type building in such areas may be unacceptable. This same issue may occur with an inland brackish groundwater desalting plant. Enhanced architectural design may be required at some additional cost to get the project permitted.

Power supply can be a cost issue. Typically, desalters, especially seawater desalters, can exert large demands on the local power grid. In the example seawater-desalting project, several sites were considered for the desalter. One of the sites would have required significant off-site power system improvements to deliver power to the plant. This would have added about 50% to the cost of the project.

Incidental Costs

Incidental costs are a necessary expense in implementing a desalting project. Engineering, environmental, legal, financial, and administrative services will be needed from the very beginning. Each project is unique and, therefore, the magnitude of the incidental costs as a percentage of the construction cost can vary over a very wide range, depending on the size (product water capacity) of the project and the project's complexity.

Typically, some percentage of the estimated construction cost is assumed for incidental costs when initially planning a desalting project. Percentages used at the planning stage for incidental costs are commonly on the order of 35%. As the project proceeds, incidental costs can be more precisely estimated and the percentage may increase or decrease.

Contingency Allowance

In the initial planning stages of a desalting project, there will be a number of "unknowns"—things that have not yet been identified. Even for "knowns," such as the desalting plant itself, unexpected costs will most likely arise as the project becomes more defined. Therefore, it is necessary to include a contingency allowance in the capital cost budget for the project. At the beginning of the project, the contingency allowance is relatively large. As the planning and design proceed, the contingency allowance may be reduced. However, it may be that some of the contingency allowance originally included in the preliminary project budget has been reallocated to estimated construction costs or incidental expenses as these costs were further defined.

The contingency allowance is typically expressed as a percentage of the estimated construction cost. In the early stages of planning, the contingency allowance may be 25% or more of the estimated construction cost. After the final plans and specifications have been prepared, the contingency allowance may be on the order of 10%. Even after construction bids are received for the project, a contingency allowance is necessary to cover unexpected costs that arise on essentially all construction projects.

ESTIMATING OPERATION AND MAINTENANCE COSTS

With respect to O&M costs, labor is typically considered a fixed O&M cost. Wage levels may change with time, but labor costs are not directly impacted by the amount of water that is desalted. Another fixed O&M cost is RO or EDR membrane replacement. Generally, once a membrane is "wetted," it is considered to have begun its useful life. That is, if a desalting plant is operated only 50% of the time, it is not generally assumed that the life of the membranes in terms of years is twice what it would be if the plant is operated 100% of the time. In addition, some percentage of the construction coswt should be included for "miscellaneous repairs and replacement" as a fixed O&M cost for planning purposes. There may be other fixed O&M costs peculiar to a specific project. Fixed O&M costs are usually measured in terms of dollars per year.

Variable O&M costs are those costs that directly depend on the amount of water that is desalted. Variable costs may include power, chemicals, concentrate disposal, and miscellaneous costs. The total annual O&M cost is the sum of the fixed and variable O&M costs in dollars per year.

Fixed O&M Costs

Labor With regard to labor, the cost is dependent on a number of factors, including the following.

- *The Size (capacity) of the Plant* A small plant may operate with only one or two permanent people,while a large plant may require operators in attendance at all times to keep up with routine maintenance, record keeping, etc.
- *The Complexity of the Plant* If the treatment process is complicated, it may be necessary to have a large operating staff to have available the needed skills.
- *Regulatory Agency Requirements* A local or state regulatory agency may have minimum staffing requirements for a desalting plant.
- *Owner's Policy* Some desalting plant owners may desire to have at least one operator at the plant around the clock, while another owner may wish to have as small an operating staff as possible with the plant operating unattended at times.
- *Local Wages/Benefits* Some areas, particularly in or near densely populated metropolitan areas, may have higher wage/benefit costs than more rural areas.

Membrane Replacement For membranes, replacement cost is usually regarded as a fixed cost because once a membrane is "wetted" it is generally assumed that membrane degradation begins. A membrane replacement fund may be established to accumulate money to replace membranes as needed. The two primary factors considered in establishing a membrane replacement fund are

- The total cost of the membranes
- The expected average life of the membranes

Determining the cost of the membranes needed in a desalter depends on

- The capacity of the desalting process (excludes desalting process by pass flow, if any)
- The recovery expected
- The blend ratio (percentage of undesalted water permissible in the product water)
- The membrane design flux (RO only)

These four factors can usually be ascertained with reasonable certainty and are usually part of a construction cost estimate.

To estimate annual membrane replacement cost, the useful life of the membranes is the only additional information needed. However, estimating the average useful life of the membranes is much more uncertain. Membrane life depends on a number of factors, at least some of which cannot be measured beforehand. Actual membrane useful life can only be determined based on operating experience. Nevertheless, an expected life can be estimated when planning a project for purposes of estimating the O&M cost and cost of water from a desalting project.

Miscellaneous   It is appropriate to include an allowance for miscellaneous O&M costs when planning a desalting project. Miscellaneous cost includes those O&M components not accounted for by the other "known" O&M cost components such as labor, power, chemicals, membrane replacement, and concentrate disposal. The miscellaneous O&M cost allowance can be estimated based on experience at other plants. However, miscellaneous O&M costs at one plant might not be truly indicative of the miscellaneous O&M costs to be realized at another plant.

Differences in design and construction, routine maintenance, climate, and so on can cause significant variations in miscellaneous O&M costs. For initial planning purposes, a percentage of the estimated construction cost is sometimes included for estimating O&M costs. For small plants with estimated construction costs of less than, say, $5,000,000, an allowance of perhaps 2% might be realistic. For a larger plant (more than $20,000,000), perhaps 1% might be used. Off-site improvements (feedwater supply and product water conveyance, for instance) associated with the desalting project should be included in the miscellaneous O&M costs.

Variable O&M Costs

Power   All of the power required by the desalting project should be included in estimating the cost of a desalting project. For instance, the power required to pump the feedwater to the plant from the source(s) of supply and the product water to customers should be accounted for as well as the power required for just the desalting plant. The power required (kW of demand and kWh consumed) can usually be relatively accurately defined in the preliminary phases of project planning.

It is important to obtain as accurate an estimate of power costs ($/kWh) as possible since power can be a substantial portion of the O&M cost and, as a result, the cost of water. This is especially true for seawater desalting plants since desalting seawater consumes more power than desalting brackish water. Most of the power cost will be a variable cost. However, some small percentage will be a fixed cost, because even when the desalter is not operating, or is operating at reduced capacity, there will be some power demand just to "keep the lights on."

Chemicals The types and amounts of chemicals that might be required in a desalting plant can vary significantly. Chemicals are required for the following purposes:

- *Filtration* If filtration is needed ahead of the desalting process, chemicals may be necessary to increase filter efficiency (suspended solids removal). Such chemicals might include chlorine, coagulants, and/or filter aides such as polymers.
- *Desalting Process Feedwater* It is common to add a scale inhibitor to desalting process feedwater to inhibit the formation of mineral scale and increase recovery. Acid may also be necessary to inhibit calcium carbonate scale formation and increase recovery.
- *Posttreatment* At the least, a disinfectant residual will probably have to be added to the desalter's product water to meet drinking water regulations. Depending on the water chemistry, addition of caustic soda to raise pH and/or lime to add hardness and alkalinity may be necessary. Use of lime in a seawater desalting plant is almost a given since the TDS of desalted seawater is primarily sodium and chloride and, consequently, it is corrosive and does not taste good.
- *Membrane Cleaning* RO and EDR membranes require cleaning at times. The frequency of cleaning and the type of chemicals needed vary depending on the particular case. Generally, a low pH (acid) solution is used to clean mineral scales and high pH (caustic soda) solution is used to clean biological fouling. In addition, a detergent may be used from time to time. If membrane filtration is used to filter the desalting process feedwater, the same chemicals may be used to clean the filtration membranes. And, in some cases, the manufacturers of membrane filters have proprietary products that they recommend for use in cleaning their membranes.

The cost for chemicals depends on the quality of the water being desalted and the product water quality goals. Chemical consumption data from a desalter treating water from the same source as a proposed desalter can be used to estimate chemical cost for a new desalter. However, if such data are not available, the cost of chemicals can be estimated based the types and amounts of chemicals from a preliminary treatment process design.

Waste Disposal In addition to the concentrate from the desalting process, a desalting plant will generate solid wastes. Disposal of the concentrate is most commonly the major concern because its volume is usually much larger than the solid waste generated and concentrate is produced continuously whenever the plant is in operation. The O&M cost for concentrate disposal can be significant. The example brackish groundwater-desalting project described previously will pay the equivalent of about $150/MG of product water delivered for concentrate disposal. While the cost of concentrate disposal in that case may have been unusually high, it is important to ascertain how concentrate will be disposed of and how much it will cost in the planning stages of a desalting project. Solid waste from a desalter may include sludge (if sedimentation/filtration is used ahead of the desalting process), cartridge filter elements, membrane elements, and other miscellaneous trash.

FINANCING COST

While the cost of financial advisors is included in incidental costs, the most significant financing-oriented cost not yet discussed is the "cost of money." In most cases, desalting projects are financed using borrowed money. The interest paid on the borrowed money increases the total cost of the project and can be significant.

To illustrate the potential significance of the cost of borrowing on the total cost of a desalting project, assume the following:

- A 2-MGD brackish water-desalting project is constructed
- Capital cost = $5,000,000
- Desalter produces 90% of its design annual yield (2 MGD × 365 days/ year × 90% = 657 MGY [2017 AFY])
- Bonds are sold to finance the project with repayment terms of
  Repaid over 20 years   Interest rate = 6%

Table 6-1 illustrates the impact of the financing cost on the cost of water ("cost of water" is defined in the next paragraph).

TABLE 6-1   Example of Impact of Financing on Cost of Water (Annual Production = 657 MGY or 2017 AFY)

	6% for 20 years	6% for 30 years
Annual payment	$436,999	$363,000
Repayment cost is equivalent to	$664/MG	$553/MG
	$216/AF	$180/AF
Total cost over repayment period	$8,720,000	$10,890,000
Bond principal	5,000,000	5,000,000
Total interest paid	$3,720,000	$5,890,000
Total volume of water produced	13,140 MG	19,710 MG
over repayment period	40,340 AF	60,510 AF
Interest paid over repayment period	$283/MG	$298/MG
is equivalent to	$92/AF	$97/AF

COST OF WATER

The sum of the annual capital repayment and annual O&M costs divided by the volume of water produced during the year is termed "the cost of water" and is usually expressed as dollars per unit volume, for example, $/MG.

TOTAL TREATED WATER COSTS

Total production costs include O&M costs and amortized capital costs and thus are dependent on the specific site and application, desalting technology used, plant facilities constructed, power, labor, and chemical costs, annual water production, and many other factors. Figure 6-2 presents typical treated water costs for seawater and brackish water desalting plants. Typical treated water costs for a 5-MGD (19,000-m³/day) seawater desalting plant range from approximately $4.00 to $6.00 per 1000 gallons; recent cost estimates for very large plants indicate that costs may be approach $3.00 per 1000 gallons for some applications. Typical brackish water total treated water costs range from about $0.60 to $3.00 per 1000 gallons for plants with 5 MGD (19,000 m³/day), depending on the site and application.

As stated previously, caution should be used when extrapolating the typical cost data presented above to a specific current or future project. They are presented only to give the reader an approximate idea of costs and detailed estimates should be made as part of any desalting plant feasibility analysis.

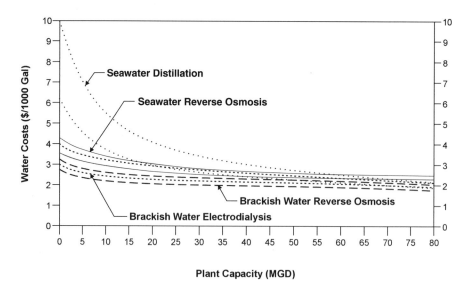

Figure 6-2 *Treated water costs for desalting water.*

SUMMARY

It is very seldom that a desalting plant is the sole facility to be constructed if desalted water is to be used as a water supply. Much more commonly, there are a number of improvements that need to be constructed away from the desalting plant site. Therefore, when considering desalting water for use as a water supply, thinking of constructing a "desalting project" rather than constructing a "desalting plant" is more conducive to including all of the costs associated with developing a salty water supply into a usable water supply.

Tables 6-2 (Estimated Capital Cost), 6-3 (Estimated O&M Costs), and 6-4 (Estimated Water Costs) present an example of how a preliminary cost estimate for a hypothetical desalting project might be presented. The example is based on

- Desalting brackish water from a surface supply (lake or river)
- Product water design capacity = 5 MGD
- Annual product water production rates
  (a) 60% of annual design capacity = 1100 MGY (3377 AFY)
  (b) 90% of annual design capacity = 1650 MGY (5066 AFY)
- Desalting process recovery = 75%
- Desalting process bypass flow (filtered water) = 1.0 MGD
- Overall recovery = 80%
- Product water = blend of 80% desalted water and 20% filtred but undesalted water

Table 6-3 illustrated how the O&M costs for the hypothetical desalting plant might be presented. Two annual production rates are given to illustrate the impact on the cost of water of spreading the fixed costs O&M over a

TABLE 6-2 Example of Preliminary Capital Cost Estimate for a Desalting Project

	Estimated Construction Cost ($)	% Construction Cost
Deliver feedwater to desalter (6.3 MGD)	2,000,000	11
Surface water filtration (6.3 MGD)	7,000,000	38
Desalting permeate capacity (4.0 MGD)	6,000,000	34
Concentrate disposal (1.3 MGD)	1,000,000	6
Product water delivery (5.0 MGD)	2,000,000	11
Estimated construction cost	18,000,000	100
Incidental costs (35%)	6,300,000	20
Contingencies (25%)	4,500,000	25
Estimated capital cost	28,800,000	145

TABLE 6-3   Example of Preliminary Estimate of O&M Costs

	Production = 1100 MGY = 3377 AFY			Production = 1650 MGY = 5066 AFY		
	Annual Cost ($)	$/MG	$/AF	Annual Cost ($)	$/MG	$/AF
Fixed O&M costs						
Labor	200,000	182	59	200,000	121	39
Membrane replacement	20,000	18	6	20,000	12	4
Miscellaneous	180,000	164	53	180,000	109	36
Subtotal	**400,000**	**364**	**118**	**400,000**	**242**	**79**
Variable O&M costs						
Power	550,000	500	163	825,000	500	163
Chemicals	150,000	136	44	225,000	136	44
Concentrate disposal	100,000	91	30	150,000	91	30
Subtotal	**800,000**	**727**	**237**	**1,200,000**	**727**	**237**
Grand total O&M	**1,200,000**	**1091**	**355**	**1,600,000**	**969**	**316**

greater volume of product water. As shown in the table, the variable O&M costs were adjusted to reflect the fact that if more product water is produced, the annual costs for chemicals, power, and concentrate disposal increase correspondingly. The annual fixed costs do not change. The result is that the O&M cost per unit of water (MG or AF) for producing 1650 MGY (5066 AFY) of water is less than for producing only 1100 MGY (3377 AFY).

Table 6-4 compares the cost of water for the two different annual production rates. The annual capital repayment cost is constant as are the fixed O&M costs. Only the variable O&M costs change with the change in annual production rate. As the "Total Cost of Water" figures in the table show, producing more water from the desalting plant would reduce the cost of water significantly.

The "Annual Capital Cost" was calculated using an interest rate of 5% and a 20-year amortization period. (Annual debt payment = 0.0802 × $28,800,000.)

However, assuming that the desalting plant's owner has more than one water supply available, the question becomes how much water should the

TABLE 6-4   Example of Preliminary Estimate of Water Cost

	Production = 1100 MGY = 3377 AFY			Production = 1650 MGY = 5066 AFY		
	Annual Cost ($)	$/MG	$/AF	Annual Cost ($)	$/MG	$/AF
Annual capital cost	2,310,000	2100	684	2,310,000	1400	456
Fixed O&M costs	400,000	364	118	400,000	242	79
Variable O&M costs	800,000	727	237	1,200,000	727	237
Total O&M costs	1,200,000	1091	355	1,600,000	969	316
Total cost of water	**3,510,000**	**3791**	**1039**		**2369**	**772**

desalting plant produce to obtain the lowest total water cost from all sources for the owner. It may be that water from the other source is less expensive than desalted water. Perhaps maximizing use of the other water source and minimizing the use of desalted water would result in a lower total water cost than maximizing the use of the desalting plant.

Chapter 7

Project Implementation

As with other types of projects, implementing a desalting project involves a number of steps. Each of these steps is described in the following pages.

DEFINING THE PROJECT

A desalting project usually consists of several distinct types of physical improvements including intake or diversion facilities from the source water, pipelines, pump stations, reservoirs, and the desalting plant. Typically, there is more than one potential site for the desalting plant. The potential for locating the desalter at any of several different sites also means that there are alternative facility configurations, including source water intake/diversion locations and design, pipeline alignments, and reservoir and pump station locations. Therefore, the first step in implementing a desalting project is the preparation of a feasibility study. The feasibility study addresses a number of issues to define project alternatives and provides an initial estimate of the capital and ongoing operation and maintenance (O&M) costs. A feasibility study typically addresses the following factors:

- Source water quantity—how much water is available to supply the proposed desalter both instantaneously (million gallons per day [MGD]) and annually (million gallons per year [MGY] or acre feet per year [AFY])
- Source water quality—physical (turbidity, suspended solids, etc.), inorganic (total dissolved solids and specific solutes, such as calcium), and

organic (any of a myriad of organic contaminants that require removal to at least some degree, including disinfection by-product precursors)

- Capacity—instantaneous (MGD) and annual (MGD or AFY)
- Product water quality—physical, inorganic, and organic
- Alternative facility locations/alignments and designs, including

    Potential source water diversion/intake points

    Alternative pipeline alignments from the potential intake/diversion points to the desalter

    Desalting plant location alternatives

    Alternative desalting plant site constraints (size, land use, power availability, access, etc.)

    Delivery of the product water to customers

    Disposal of treatment process residuals, especially the concentrate, from the desalting process

    Environmental and permitting considerations (see below)

    Other site-specific considerations peculiar to the project

    Preliminary estimates of the capital and operation and maintenance costs for the alternatives

PUBLIC INVOLVEMENT

It is becoming more and more common for construction projects to be scrutinized by the public. This is especially true for what might be considered controversial projects. Desalting projects are often considered controversial because of the nature of the project (converting undrinkable water to drinkable water), the location of the facilities (adjacent to the ocean for seawater desalting projects, for example), and the probably higher cost of the desalted water as compared to existing, "conventional" water supplies.

Including a public education/information program as part of desalting project should be considered. The scope of the program needs to be tailored to the specific project and the perceived public concerns. It is probable that during the course of the project that modifications will be needed to the initial concept of the public education/information program.

ENVIRONMENTAL DOCUMENTATION

Environmental concerns must be addressed from the beginning of project planning. Some potential desalting plant sites may be unacceptable because of environmental issues. In many cases, an environmental problem for a specific site can be readily determined in the early stages of project planning, thereby eliminating that site from further consideration.

Input from a design professional is essential for preparation of the environmental documentation for a project. A design professional must provide information on the design and operation of the proposed desalting project so that appropriate environmental documentation, including environmental mitigation measures, can be prepared. Environmental issues may be the determining factor in selecting which of several alternative desalting project options may be implemented at the least cost.

PERMITTING

Permitting and environmental issues are closely linked. Obtaining at least some of the permits that will be required to construct and operate a desalting project will most likely be dependent on the results of environmental studies. This is especially true when addressing desalting process concentrate (brine) disposal. As with environmental documentation, input from a design professional is necessary during the permitting process.

FINANCING

Unless the project owner has sufficient cash reserves, grants, or other sources of funding available to fund the project, debt financing of the capital cost will be necessary. The project financing package may include interest during construction and, perhaps, sufficient funds to make the initial debt service payments. In some cases, the first year's O&M costs have been included in the debt financing to provide time for revenues from the sale of the desalted water to be received. One financing option available in some situations is private financing, as discussed further on in this chapter.

DESIGN

A preliminary design report (PDR) should be prepared for the selected project alternative. A PDR provides more detail on the design of the proposed project. The PDR should also address environmental mitigation and permitting compliance. Care must be taken in preparing the PDR and the environmental documentation so that the environmental documents can withstand legal challenges. Waiting until final design of the project to address permitting and environmental issues is too late in the project implementation process. Significant changes from the project as described in the environmental documentation or in permits can lead to challenges and delays, or cancellation, of a project because the final project design is too different from the design discussed in the PDR. Final design of the project facilities is most efficiently accomplished after environmental mitigation measures and permit require-ments affecting the final design have been thoroughly reviewed and defined.

OBTAINING CONSTRUCTION BIDS

Depending on how construction bids are solicited, the bid documents can be either

- 100% complete design drawings and specifications, or
- Less than 100% complete plans and specifications.

The bid solicitation procedures in which 100% complete and less than 100% complete design plans and specifications are used are discussed further on. (See **Project Implementation Alternatives.**)

PREPURCHASING OF EQUIPMENT

In addition to soliciting bids for constructing the desalting project, separate contracts can be awarded for purchase of specific equipment. Desalting equip-ment is a candidate for prepurchase by the owner and installation by the construction contractor. Prepurchase of desalting equipment also provides the opportunity to complete the final design of the desalting plant based on the particular desalting equipment selected for the project thereby making the final design of the desalter more efficient and cost-effective. Prepurchasing the desalting equipment may also accelerate the project schedule because fabrication of the desalting equipment can begin earlier in the project imple-mentation process. The equipment prepurchase and construction contract bid documents can be written so that the construction contractor is responsible for installing the owner-furnished equipment.

Potential advantages that may result from prepurchasing equipment include

- Acceleration of the project schedule
- Reduction of the cost of the equipment by avoiding the construction contractor's markup
- More owner control over the design and quality of the equipment pur-chased

Potential disadvantages to the project owner associated with prepurchasing equipment include

- Added contract administration responsibility
- Coordination between the prepurchased equipment supplier and the construction contractor
- Correction of problems with or caused by the owner-furnished equipment

CONSTRUCTION

The project owner's participation in the construction phase of the project can range from providing full-time on-site construction management, including inspection, to near total reliance on the construction contractor. (See **Project Implementation Alternatives.**) One option for a project's owner to consider is to have the owner's operating staff involved during construction of the desalting plant. Doing this provides a valuable opportunity for the owner's operators to become familiar with the plant's design features. This usually leads to fewer operating problems during start-up. However, caution is needed so that the operating personnel do not unilaterally make design changes that will impact the construction contractor's costs and/or the completed project's design and operation.

One particular aspect of a desalting plant that merits special attention during construction is the instrumentation/control system. The level of instrumentation/control schemes typically employed in modern desalting plants is complex. Debugging the instrumentation/control system usually takes some time and is best performed by the firm that provided the instrumentation/control equipment working in close cooperation with the professionals who designed the instrumentation/control system.

OPERATING A DESALTER

Operating a desalting project requires personnel with particular skills. A detailed and thorough operator-training program is needed. As noted above, the instrumentation/control system for desalters is usually complex. It is best if the firm that supplies the instrumentation/control system provide training for the plant's operators. Similarly, the vendor who provided the desalting equipment should provide start-up and operator training services as part of their contract. The equipment vendors should also provide detailed operation and maintenance manuals for their equipment. These manuals should be used in the training sessions mentioned above.

In addition to the initial operator training, requiring the equipment vendors to provide one or more additional training sessions sometime after the plant has started operation is beneficial. The operators will encounter situations not specifically addressed in the initial training sessions. Operators may also suggest changes in the operation of the plant after they become familiar with the plant. Any such changes should be discussed with the plant's designers and equipment suppliers before implementing the changes because of possible impacts on other plant equipment or processes.

The desalter's designers should prepare an operations manual describing how the plant was designed to be operated. The operations manual should address each unit process describing

- Its purpose
- How it functions
- Instrumentation/control devices and logic
- Troubleshooting

However, assistance with the preparation of an O&M manual and operator training should be included in the equipment prepurchase and/or construction bid documents. While some parts of a desalter may be typical of desalters in general, there are substantial and, sometimes, subtle differences between desalter designs. An operations manual must be specific to the plant for which it is written. Adequate funding of operations manual preparation and operator training is imperative to achieving the least practical O&M costs, reliable operation, and consistent product water quality.

OBTAINING THE REQUIRED EXPERTISE

To accomplish the desalting project implementation steps described above, several types of expertise are needed, including

- Engineers from a number of disciplines—civil, process, structural, electrical, mechanical, instrumentation/control, geotechnical
- Public information/education
- Architects
- Environmental specialists
- Permitting specialists
- Financial experts
- Attorneys
- Equipment supplier(s), especially for specialized equipment such as desalting apparatus, to provide equipment and assist in the start-up
- Construction contractor(s)

Large entities contemplating construction of a desalting plant may have personnel with some or all of the different types of skills and experience needed in-house with the probable exception of the equipment suppliers and construction contractors. However, in-house personnel may have other assignments that preclude them from working on the desalting project. If in-house personnel do not have the experience and/or the time to devote to the project, retaining the services of consultants to provide skills and expertise necessary is a commonly employed option.

PROJECT DELIVERY ALTERNATIVES

There are several ways in which a project's owner can accomplish the steps described above, including the following:

- Design/bid/build
- Design/build
- Design/build/operate
- Design/build/operate/finance
- Design/build/own/operate/transfer

These project implementation schemes, and variations on them, are discussed in the following paragraphs.

Design/Bid/Build (DBB)

DBB is the conventional means of constructing most municipal projects and many privately owned projects. A detailed set of construction bid documents (plans and specifications plus other documents that impact the design and/or construction) are prepared by a design professional. Construction bids are then solicited. A contract is then awarded, generally to the lowest bidder, to construct the project.

In a DBB project, the project's owner obtains the services of engineers, architects, environmentalists, permitting specialists, financial experts, and attorneys. Some or all of these services may be available in-house or from consulting firms. These people work together as a team to plan, prepare environmental documentation, arrange financing, obtain permits, design, and prepare construction plans and specifications for soliciting construction (and equipment) bids.

This is the easiest type of project for construction contractors to bid. With a well-prepared and thought out set of plans and specifications upon which to bid, there is very little guesswork required on the part of the construction contract bidders as to what will be required of them. Substantially all of the environmental and permitting work has been completed. Environmental and permit compliance requirements are included in the bid documents. Rights-of-way for constructing the project have been acquired by the project's owner. The owner has also arranged for financing. The construction contractor's responsibility is limited to constructing the project according to the plans and specifications prepared by the project's owner.

Design/Build (DB)

With the design/build approach, the owner issues requests for proposals to design and build the project. The owner arranges for financing and pays the DB contractor for performing the work. The contents of the request for proposals can include as much or as little information as the owner wishes to provide. For example, in addition to designing and constructing the project, the project's owner may leave environmental and permitting work up to the contractor. The contractor may even be required to acquire rights-of-way.

It has been suggested that several benefits accrue to the project's owner by using DB rather than the more conventional DBB approach. The claimed benefits include

- Quicker completion of the project
- Lower overall project cost
- Innovative design ideas that might not be included by the designer in a DBB project

Essentially, the work to be accomplished in either the DBB or the DB approaches is the same. The difference between these two is how much of the work (responsibility) will be transferred to the contractor to accomplish. Whether any or all of the suggested DB benefits will accrue to the benefit of a project's owner for a specific project is questionable and depends greatly on the level of risk assigned to the contractor by the owner. For example, if the contractor is assigned the tasks of environmental documentation and obtaining permits, there is, at least in most cases, no reason to believe that a private contractor can complete either of the these tasks more expeditiously than the owner. The acceptability of environmental documentation is subject to the scrutiny of any number of groups private and public. The issuance of permits is typically under the control of myriad public agencies and regulatory bodies. And, in many cases permits will be issued only to the project's owner.

The costs associated with environmental compliance and obtaining permits can be substantial. There may be permit costs levied by the regulatory agencies that cannot be determined during the bidding period. And environmental mitigation measures that may be imposed on the construction contractor can greatly increase the cost of construction. If environmental mitigation measures that affect construction are not known during the bidding period but have to be estimated it is highly likely that the bidding contractor's will choose to err on the side of conservatism. That is, bidders will include contingencies in their bids for designing and building the project.

With respect to obtaining rights-of-way, comments similar to those made regarding environmental documentation/compliance and permitting apply.

Rights-of-way and/or encroachment permits may be needed from govern-mental agencies also. Obtaining these takes time and costs money. There may also be conditions included with the rights-of-way and/or encroachment per-mits that impact the cost of construction.

Requiring the DB contractor to take on responsibilities (liabilities) can be done for a cost. Passing responsibilities onto the DB contractor may be lik-ened to buying insurance on a home or car—the lower the deductible, the higher the insurance premium. The greater the risk the DB contractor is re-quired to assume, the higher the contractor's bid for the work.

Another factor that project owners thinking of DB should consider is that somebody has to prepare plans and specifications for the project. If the owner has this done, the owner knows what he or she will have when construction is complete. If the owner prepares the plans and specifications, as in a DBB project, all of the equipment, materials, layout, and so on will be defined prior to bidding. The owner will know the project's design before it is built. With DB, the project's owner will have only limited input into the design of the project, depending on how the DB bid documents are written. The more complete the DB bid documents are, the more the project's owner will have to say about the design of the project.

However, one of the benefits suggested for DB is innovative design and/or construction ideas from the DB team. The more complete the DB bid documents, the less opportunity there is for the DB contractor to innovate. Therefore, a careful balance between the completeness of the bid documents and the owner's desire to encourage innovation on the part of DB bidders is needed.

With respect to innovation, there is one other consideration. Innovation, by its very nature, implies risk. Innovation is the employment of new ideas, materials, methods of construction, and so on in an attempt to reduce costs and/or time required for construction or to obtain a better constructed project as compared to more conventional designs and/or construction methods. The primary objective of DB bidders is to make a profit. If an innovative idea should be used in the design or construction of a project and not be as suc-cessful as hoped, the DB contractor's expected profit will be adversely im-pacted. One means of encouraging innovation is for the owner to share in the risk of employing innovative designs, equipment, and/or construction means for a project. If a project owner should decide to use the DB approach, the owner should carefully consider employing knowledgeable and experienced personnel to represent the owner's interest in the design and construction of the project.

Design/Build/Operate (DBO)

DBO is like the DB approach except that the contractor remains on site to operate and maintain (O&M) the desalter after construction is completed. The DBO contractors' bids include costs to provide O&M services for the desalter for some period of years. The DBO bid documents can require that bidders include labor, parts replacement, maintenance, "consumables," insurance, and so on in their bid for the project. Consumables include things used in operating the plant, such as chemicals, power, and RO membrane elements. For power and chemicals, it is preferable to request DBO bidders to provide maximum guaranteed consumption rates (pounds of chemical per MG of water produced, kWh per MG, etc.) rather than costs for these items. This is because different bidders may use different unit prices for chemicals and power, thus making evaluation of the bids more difficult. The prices for power and chemicals can change dramatically, too, especially over the period of years for which the owner may wish the DBO contractor to operate the plant. For RO plants, the RO membrane supplier can be asked to provide maximum annual membrane replacement cost guarantees over the life of the O&M contract. Typically, O&M contract costs are tied to some inflation-reflecting index to allow for cost increases due to increases in labor rates and other factors.

It is common to compare DBO bids based on a life-cycle cost analysis. The lowest construction cost bid may not be the most economical bid. The cost of water over the life of the project for the several bids received can be calculated to see which proposal would result in the lowest cost over the period of time the bids are compared.

Advantages to the owner include

- The DBO contractor is responsible for operating and maintaining the facility he has designed and built, thus relieving the owner from hiring and training operators.
- The O&M cost the owner needs to budget for is reasonably well known over the period for which the DBO contractor will operate and maintain the desalter.
- If the desalter should not operate as well as guaranteed by the DBO contractor, the DBO contractor is responsible for correcting any deficiencies in the design or construction of the desalter.

The caveats mentioned in the discussion on DB apply to DBO also.

Design/Build/Operate/Finance (DBOF)

DBOF is similar to DBO except that the responsibility for financing the project is also transferred from the owner to the contractor. If this approach is used to implement a desalting project, the resulting contract between the owner and the DBOF contractor may be considered to be a service contract; that is, the owner will be buying from the DBOF contractor some amount of water at some cost over some period of time. An important consideration in this approach is the cost of financing. Generally, the cost of financing for a private entity is more than the cost of financing for a public entity. If possible, arranging for public (tax-exempt) financing is preferable to reduce interest costs. The issues mentioned in the preceding discussions of DB and DBO apply to the DBOF approach also.

Design/Build/Operate/Own (DBOO)

In this case, the DBOO contractor, in addition to designing, building, and operating the desalter, would also own the project. Implied is that the DBOO contractor would also finance the cost of the project. After some period of time has elapsed, ownership of the project may be transferred to a new owner such as a municipal agency. At that time, the new owner either takes responsibility for O&M of the project or contracts out these services. The issues discussed for DB, DBO, and DBOF also apply in this case.

COMPARISON OF DESALTER IMPLEMENTATION ALTERNATIVES

Table 7-1 compares the contractor's responsibilities for the five different project implementation schemes discussed in this chapter. The primary difference between the five schemes is allocation of risk. For the DBB approach, the project's owner takes responsibility for the design, operation, and financing of the project, including meeting environmental requirements and obtaining necessary permits. The contractor's obligation is limited to constructing a desalter that meets the requirements of the construction bid documents (plans and specifications).

By using the DB approach, the owner transfers the risk of design and, possibly, the risk for obtaining environmental and permitting approvals to the DB contractor. This transfer of risk from the project's owner to the DB contractor comes at some cost, however. Sharing of risks by the owner and the DB contractor may be the least costly means of implementing the project. Determining the allocation of risks between the owner and the DB contractor can require extensive negotiations.

DBO can transfer even more risk from the owner to the DBO contractor. In addition to having to design and build the project, the DBO contractor

TABLE 7-1 Comparison of Contractor's Responsibilities for Project Implementation Schemes

	Design	Build	Operate	Finance	Own
Design/bid/build		•			
Design/build	•	•			
Design/build/operate	•	•	•		
Design/build/operate/finance	•	•	•	•	
Design/build/operate/own	•	•	•	•	•

must operate and maintain the desalter for some period of years. The water produced by the desalter must meet all applicable water quality regulations. And water quality regulations are constantly changing and, in general, becoming more stringent with time. As with DBO, allocation of risk between the owner and DBO contractor can be the subject of extensive negotiations, especially with respect to the costs associated with risk assumption.

Adding financing (DBOF) to the contractor's responsibilities transfers yet more risk from the owner to the contractor. Again, this transfer of risks is subject to negotiation and cost considerations. In the case of a publicly owned project it may be less expensive for the public agency to finance the project than for a private contractor to do so, at least when considering long-term financing as opposed to short-term construction period financing.

Finally, essentially all risks associated with the project can be transferred to the DBOO contractor. The only responsibility of the entity receiving desalted water from the project is to take and pay for the water according to the terms of the contract between the water purchaser and the DBOO contractor.

All of these methods, and even some variations on them, have been used to implement desalting projects. Care must be taken in selecting the most appropriate scheme for a specific project. Knowledgeable and experienced persons should be employed to advise any entity contemplating a desalting project. And remember that transfer of risks from the beneficiary of the desalting project (the entity receiving the desalter water) to another entity (private contractor) can be achieved only at some cost.

Lavaca-Navidad River Authority Seawater Reverse Osmosis Facility

The Lavaca-Navidad River Authority is contemplating the construction of a desalting plant to produce drinking quality water for water short planning areas. The recent technological advances in desalination coupled with ongoing power deregulation of the electric power generating industry provide a "window" during which the feasibility of desalination opportunities may be investigated. One example of the reduction in the cost of water is the recent Tampa Bay Desalination Project. This facility will produce fresh drinking water for the Tampa Bay area at an initial cost of $1.71 per thousand gallons and an average cost over the life of the facility of approximately $2.11 per thousand gallons.

The technology to remove dissolved salts from bay water is currently available and used in many locations throughout the world. The technology is effective, reliable, and capable of providing large quantities of freshwater from seawater and brackish water sources. The deterrent to wider usage has been cost. However, recent advances in the removal of dissolved salts have resulted in drastic reduction in the cost of water from such water treatment processes. This is true particularly at favorable locations.

PURPOSE

The purpose of this work was to determine the most cost-effective, environmentally responsible process for desalting seawater to provide a drought-proof water supply for users in San Antonio and/or Corpus Christi areas. The alternative under study involves the co-siting of a desalination plant with Central Power & Light's (CP&L) Joslin Power Station in Point Comfort, Texas.

This report presents the preliminary technical and economical results of the construction of a 50,000- to 100,000-acre-feet per year (AFY) desalting facility for the Lavaca-Navidad River Authority. The plant would be located at the CP&L Joslin Power Station in Point Comfort, Texas. The intent of locating the desalination plant at this site is to determine if a dual-purpose facility would offer an economic benefit, that is, a facility that produces both power and water.

A dual-purpose facility is one in which two processes (power and water) share the use of steam. This has the advantage of lowering the fuel usage. This benefit, however, can be obtained only if the power plant has been designed initially to feed steam to a desalting plant. This, of course, is not the case at the CP&L station. The existing power plant has been designed as a single-purpose installation (i.e., the production of power only). Although the existing power plant cannot be used for this purpose, other inherent benefits can be realized. These include the following:

- Sharing of the existing seawater intake system
- Sharing of operating and maintenance staff
- Sharing of maintenance facilities
- Land purchase is not required

To take advantage of these benefits, this study assumes the construction of a dual-purpose facility at this site. Two desalting processes were considered:

- Seawater reverse osmosis (SWRO)
- Multiple-effect distillation (MED)

TECHNICAL DISCUSSION

When the fuel usage of the dual-purpose arrangement is compared with the fuel usage from stand-alone power and desalting arrangements, it is found that the dual-purpose arrangement has a much lower fuel consumption. This is the advantage of the dual-purpose facility. This fuel reduction can be as much as 60–70%. Therefore, this process arrangement is often chosen for the construction of desalination processes (i.e., multistage flash, multiple-effect distillation, and vapor compression). These processes use steam as the heat source and therefore lend themselves to the dual-purpose arrangement. However, because the RO process uses significantly less energy than all the thermal systems, cost comparisons must be made between the stand-alone RO process and the dual-purpose installations.

For the RO process, the water quality to be treated is the key consideration that determines the design and ultimate cost of water. That is, the lower the total dissolved solids (TDS) of the supply to be treated, the lower the energy usage of the process. The water to be treated at the Point Comfort site has a TDS of much lower than standard seawater. Table 1-1 gives the assumed concentration of dissolved solids in the water to be treated. This is the 90th percentile concentration and this value was used to present the costs that would be appropriate during the summer under low-flow and high-demand conditions. The average concentration of dissolved solids at this location is approximately 14,000 mg/L, based on data collected both before and after the installation of Lake Texana upstream of the bay that the cooling water for the power plant is drawn.

Table 1-1, gives the assumed water quality of this study. This shows that the TDS is about 25,000 mg/L—much lower than that of standard seawater, which has a typical TDS of 34,500 mg/L. Thus, for the RO process, the operating pressure of the high-pressure pump (HP pump) will be much lower than that for treating standard seawater. Also, because of the lower TDS, the process recovery will be much higher. These factors combine to result in a very cost-effective design.

To determine if the dual-purpose plant offers an economical benefit, a base case is first presented. This case is a stand-alone RO plant that uses motor drives for the HP pumps. This case is then compared with two dual-purpose arrangements:

- An RO process using steam turbine drives for the HP pumps
- A hybrid process using RO and multiple-effect distillation (MED)

Two production rates are studied, one for a yearly production of 50,000 AFY and one for 100,000 AFY. Of course, there is a major benefit of constructing

TABLE 1-1 Feedwater Quality

Constituent	Concentration (mg/L)
Anions	
Calcium	290.8
Magnesium	924.7
Potassium	276.3
Sodium	7,674.2
Strontium	9.5
Cations	
Bicarbonate	101.8
Chloride	13,798.5
Fluoride	0.0
Nitrate	0.0
Sulfate	1,925.8
Phosphate	0.0
Total dissolved solids	25,001.6
pH	8.2
Temperature (°F)	90.0

the dual-purpose facility—the production of electricity—but we do not address that in this presentation.

Base Case—SWRO, Variable Speed Drives for HP Pumps

To compare the benefits these processes may offer, a base case is first presented. This is a stand-alone, single-purpose RO plant that uses motor drives for the HP pumps. The base case RO process characteristics are given in Table 1-2 and the process diagrammatic in Figure 1-1. This case assumes a stand-alone RO plant that uses variable speed drives for the HP pumps. Electricity is purchased from the grid. To meet the required productions, it is assumed that each train will be sized at 5.0 MGD. Thus, 11 trains are required to meet the production of 50,000 AFY, 22 trains will be required to meet the production rate of 100,000 AFY.

Case 1—SWRO, Steam-Driven HP Pumps

The process for case 1 is RO, which uses a steam turbine for driving the HP pump, taking steam from the power plant. The process characteristics of this design are the same as that for the base case with the exception that the steam requirement from the power plant would be 458,890 lb/h for the production rate of 50,000 AFY and 917,774 lb/h for the production rate of 100,000 AFY.

TABLE 1-2   Technical Characteristics Base Case—Stand-Alone SWRO

	Per Train	50,000 (AFY)	100,000 (AFY)
Flow rate (gpm)			
Seawater supply	6,944	76,384	152,768
Permeate	3,472	38,194	76,384
Concentrate	3,472	38,194	76,384
Pressures (psig)			
Seawater supply	10	10	10
High-pressure pump	758	758	758
Permeate	15	15	15
Concentrate	753	753	753
Seawater supply temperature (°F)	90	90	90
Connected electrical load (kW)	4,400	48,400	96,800
Specific energy consumption (kWh/kgal)	12.9	12.9	12.9
Water quality (mg/L)			
Seawater supply	25,001	25,001	25,001
Permeate	283	283	283
Concentrate	49,719	49,719	49,719
Recovery (%)	50	50	50
Number of units	1	11	22
Production capacity (MGD)	5	55	110

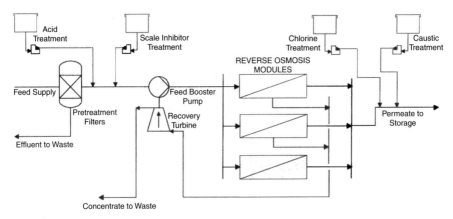

Figure 1-1   *Lavaca-Navidad Seawater Reverse Osmosis Facility—base case.*

Figure 1-2 gives the process diagrammatic for this case. The process characteristics can be found in Table 1-3.

Case 2—Hybrid SWRO and MED Design

Case 2 considers using a hybrid plant that combines the RO and MED processes. The advantage gained by this arrangement is that the steam from the power plant is first used to power the high-pressure pumps in the process. The exhaust from these turbines is then used as the heat source for the MED process. Thus, the cost of steam is attributed to only the RO process and not the MED process. The production of water from this system is about 80% from the RO process and 20% from the MED process. This case would re-

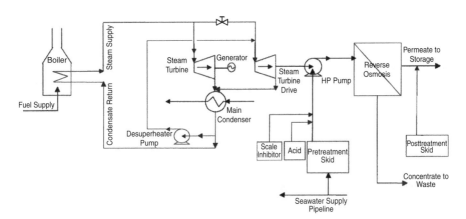

Figure 1-2   *Lavaca-Navidad Seawater Reverse Osmosis Facility—case 1.*

TABLE 1-3   Technical Characteristics Case 1—Dual-Purpose SWRO, Steam-Driven Pumps

	Per Train	50,000 (AFY)	100,000 (AFY)
Flowrate (gpm)			
Seawater supply	6,944	76,384	152,768
Permeate	3,472	38,194	76,384
Concentrate	3,472	38,194	76,384
Steam flow (lb/h)	41,717	458,890	917,774
Pressures (psig)			
Seawater supply	10	10	10
High-pressure pump	758	758	758
Permeate	15	15	15
Concentrate	753	753	753
Steam pressure	250	250	250
Seawater supply temperature (°F)	90	90	90
Connected electrical load (kW)	4,400	48,400	48,400
Specific energy consumption			
steam drive (kWh/kgal)	3.39	3.39	3.39
Water quality (mg/L)			
Seawater supply	25,001	25,001	25,001
Permeate	283	283	283
Concentrate	49,719	49,719	49,719
Recovery (%)	50	50	50
Number of units	1	11	22
Production capacity (MGD)	5	55	110

quire approximately nine trains of RO and two trains of MED for the 50,000 AFY case and 18 trains of RO and 4 trains of MED for the 100,000 AFY case. The process characteristics can be found in Table 1-4. This process diagrammatic is shown in Figure 1-3.

ECONOMICS

The costs presented herein are based on a proprietary desalting cost program developed by DSS Consulting, Inc. over the past twenty years. This program is continually updated based on desalting plant bids and vendor quotes.

Cost Basis

The following assumptions were made in the preparation of these costs:

Plant location   Joslin Power Station
Cost year   Mid-2000
Service life   30 years
Interest   6%

**TABLE 1-4   Technical Characteristics Case 2—Dual-Purpose Hybrid Design
(50,000 AFY)**

	RO per Train	MED per Train	RO Total	MED Total	Total
Flow rate (gpm)					
Seawater supply	6,944	7,434	62,996	14,868	77,364
Permeate	3,472	3,472	31,248	6,944	38,192
Concentrate	3,472	3,962	31,248	7,924	39,172
Steam supply	37,749	37,749	339,739	339,739	339,739
Pressures (psig)					
Seawater supply	10	10	10	10	—
High-pressure pump	758	—	758	—	—
Permeate	15	0	15	0	—
Concentrate	753	0	753	0	—
Seawater supply temperature (°F)	90	90	90	90	90
Connected electrical load (kW)	4,400	1,129	39,600	2,258	41,858
Specific energy consumption (kWh/kgal)	3.38	5.42	3.38	5.42	3.75
Water quality (mg/L)					
Seawater supply	25,000	25,000	25,000	25,000	25,000
Permeate	283	25.0	283	25.0	236
Concentrate	49,641	46,886	49,641	46,886	49,083
Recovery (%)	50	46.7	50	46.7	49.4
Number of units			9	2	11
Production capacity (MGD)	5	5	45	10	55

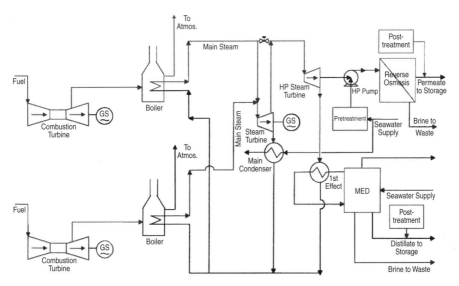

Figure 1-3 *Lavaca-Navidad Seawater Reverse Osmosis Facility—case 2.*

Purchased electricity cost	$0.04/kWh
Joslin Power Station electricity cost	$0.0381/kWh
Steam cost	Varies per case
Land cost	$0.00
Direct labor cost	$12.00/hour (average all disciplines)
Labor overhead	40% of direct labor
Chemical costs	
Chlorine	$0.20/lb
Sodium bisulfite	$0.88/lb
Scale inhibitor	$0.87/lb
Coagulant aid	$0.10/lb
Cleaning chemicals	$0.008/kgal
Construction overhead and profit	8% of total construction costs
Owners' costs	7.5% of total construction costs
Contingency	7.5% of total construction costs
Freight and insurance	5% of total construction costs
Taxes	None

Reverse Osmosis Process

The costs presented herein include the following equipment for this process:

- Feedwater supply piping
- Interconnecting feedwater supply piping and valves
- Coarse and dual media filtration pretreatment
- Chemical pretreatment systems
- Cartridge filters
- Process equipment, including all pumping services, energy recovery equipment, membrane cleaning equipment, electrical distribution and instrumentation, and controls
- Building to house the control room, process equipment, chemical treatment, motor control center, and offices
- Chemical posttreatment equipment
- Emergency generator
- Main step-down transformers
- Site development
- Concentrate disposal system

Multiple-Effect Distillation

The costs presented herein include the following equipment for this process:

- Feedwater supply pumping
- Interconnecting feedwater supply piping and valves
- Chemical pretreatment equipment
- Process equipment, including all pumping services, brine heaters, heat rejection systems, electrical distribution, and instrumentation and controls
- Concentrate disposal system

The capital and operating costs for each case are given in Table 1-5.

ENVIRONMENTAL CONSIDERATIONS

A review of the potential environmental impacts of the proposed project was performed as a part of this study. While the discharge of the plant concentrate and suspended solids load back to the bay in the cooling water discharge line is the least expensive option for disposal, this may not be feasible from an environmental standpoint. The environmental issues investigated did not uncover a "fatal flaw" that would prohibit consideration of the process, but there is the possibility of increased costs for disposal of the hypersaline reject water and the solids (mud, silt, etc.) that would be removed by the pretreatment process. These disposal costs could add an additional $0.25 per 1000 gallons of treated water produced to the cost of the process. In this study the costs for the pretreatment of the water were based on a conventional pretreatment scheme that used chemical addition. In fact, an alternative scheme was also investigated that is less expensive and does not add chemicals to the solids, making it more likely that the solids could be returned to the environment. Any cost reductions from using this alternative process will be determined during further studies.

TABLE 1-5 Capital and Operating Costs

Case	Capital Cost ($)	Operating Cost ($/kgal)	Fixed Cost ($/kgal)	Cost of Water (k/gal)
First-Year Costs—50,000 AFY				
Base case (stand-alone RO)	160,906,500	1.24	0.65	1.89
Case 1 (DP steam drives)	160,815,000	1.64	0.71	2.35
Case 2 (DP hybrid)	170,361,500	1.47	0.76	2.23
First-Year Costs—100,000 AFY				
Base case (stand-alone RO)	195,695,000	1.15	0.43	1.58
Case 1 (DP steam drive)	205,698,000	1.59	0.46	2.04
Case 2 (DP hybrid)	258,686,000	1.41	0.57	1.98

CONCLUSIONS

The following conclusions can be drawn from this study:

- The most cost-effective process is the stand-alone RO system. This process exhibits the lowest capital and operating cost and the lowest cost of water.
- Locating this process at the Joslin Power Station offers the following advantages:
  a. The seawater intake exists and thus a new intake would not be required.
  b. Sharing of operating staff.
  c. Sharing of maintenance staff.
  d. Sharing of workshop.
  e. Land for locating the plant would not have to be purchased. This assumes that the Joslin plant is purchased by a party that is amenable to the installation of the desalination plant on site as a viable and valuable base load customer to the power plant.
  f. The concentrate discharge from the desalting plant can use the existing cooling water outlet from the power plant, provided that the environmental issues can be resolved. If that is not the case, then there would be an additional cost for handling of the concentrate and solids.
  g. The diversion of the treated water portion of the cooling water effluent from the power plant reduces the heat load on the receiving waters at the cooling water discharge. In addition, if the concentrate from the desalination plant can be mixed with the remaining cooling water discharged to the bay, there will be a further reduction in the heat load returning to the bay.

Two Years of Operating Experience at the Port Hueneme Brackish Water Reclamation Demonstration Facility

Todd K. Reynolds

Kennedy/Jenks Consultants, San Francisco, California

Frank Leitz

U.S. Bureau of Reclamation, Denver, Colorado

In Port Hueneme, California, a state-of-the-art desalination facility (Figure 2-1) uses three brackish water desalination technologies: reverse osmosis (RO), nanofiltration (NF), and electrodialysis reversal (EDR), operated side-by-side to produce up to 4 million gallons per day (MGD) of high-quality drinking water. The Brackish Water Reclamation Demonstration Facility (BWRDF) is the cornerstone of the Port Hueneme Water Agency's (PHWA) Water Quality Improvement Program. In addition to providing desalted water for local use, the BWRDF also serves as a full-scale research and demonstration facility.

It is usually a difficult task to compare the long-term performance and operating costs of different technologies due to variables in source water quality, plant capacities, and labor, power, and chemical costs. Operating three full-scale desalination technologies in parallel at the same site and on the

Figure 2-1   *The desalination facility in Port Hueneme, California.*

same source water has made direct comparison possible. During the course of the plant's operation, the PHWA staff collects data on operating costs and performance characteristics of the three membrane systems. This provides a basis for the comparison of these three desalination technologies that can then be used by water purveyors to determine which technology best suits their local conditions.

This Case Study provides an overview of the PHWA BWRDF, highlights some lessons learned after two years of operation, and presents a comparison of the three desalination technologies. Over two years of operation the BWRDF produced greater than 3 MGD of high-quality drinking water. In the first year, the facility experience problems with biofouling in the RO and NF membranes and the EDR system had the lowest operating and maintenance (O&M) cost at $0.34 per 1000 gallons (kgal). In the second year, operation changes were made to resolve the biofouling and now the RO and NF system have the lowest O&M cost at $0.31/kgal.

BACKGROUND

Prior to the implementation of the PHWA's Water Quality Improvement Program, PHWA's customers (City of Port Hueneme, Channel Islands Beach Community Services District, Naval Construction Battalion Center–Port Hueneme, and Naval Air Weapons Station–Point Mugu) were concerned with the long-term reliability and water quality of their existing water supplies. Each of these water purveyors utilized brackish groundwater from the Oxnard Plain Groundwater Basin, which is a critically overdrafted basin and is under active basin management. The groundwater management plan calls for reductions in groundwater extractions by 25% over a 25-year period. Until the early 1990s, groundwater had been extracted from local deep aquifer wells along the coast that were increasingly subject to seawater intrusion. The United Water Conservation District (UWCD) supplied additional water from inland shallow aquifer wells. The total dissolved solids level of these water sources is normally greater than 1000 mg/L. Although the groundwater meets the California Department of Health Services' primary drinking water standards, it is highly mineralized and is aesthetically unpalatable. Furthermore, customers were burdened with added costs for water softeners and bottled water, as well as the indirect costs, including shortened plumbing and appliance life and staining on glassware and laundry.

In response to increasing overdraft of the local groundwater basin, seawater intrusion, and poor groundwater quality, PHWA implemented a Water Quality Improvement Program. The program includes demineralizing the local shallow groundwater, which is used in conjunction with imported surface water from the California State Water Project.

PROJECT BENEFITS

Implementation of this innovative program has provided significant benefits to over 55,000 people in southwest Ventura County, California. Benefits include improved water quality, joint use of facilities, and obtaining a long-term, safe, reliable, and environmentally sustainable high-quality water supply that meets current and proposed drinking water quality standards under the Safe Drinking Water Act. Shared use of facilities by PHWAs customers has reduced capital and operating costs for the four wholesale customer agencies. Long-term water supply reliability for the PHWAs customers has been improved by access to both demineralized groundwater from local sources and imported State Water Project water. The delivery of imported State Water Project water has also allowed the PHWAs customers to reduce groundwater extractions from coastal wells threatened by seawater intrusion. Relocating groundwater extractions from the coastal area to inland recharge areas also minimizes seawater intrusion. In addition, demineralizing the water prior to distribution reduces the need for point-of-use water softening and/or reverse osmosis units.

PROJECT TIMELINE AND BWRDF COST

The PHWAs Water Quality Improvement Program was implemented over a 6-year period starting in 1993 and culminating with the startup of the BWRDF in 1999. The BWRDF has been in continuous operation since January 1999. The cost of the BWRDF was $5.7 million. The cost of PHWA Water Quality Improvement Program also included approximately $7 million for raw and treated water pipelines. The United States Bureau of Reclamation provided 25% of the facility cost because the BWRDF design permits side-by-side comparison of RO, NF, and EDR desalination technologies and serves as a full-scale brackish membrane research and demonstration facility.

BWRDF OVERVIEW

The initial design is a three-membrane treatment processes. RO, NF, and EDR operate side by side to produce a total of 3 MGD of treated and blended water, as shown in Figure 2-2.

Pretreatment

The source water for the BWRDF is chloraminated groundwater from inland, shallow aquifer wells operated by the UWCD. The shallow aquifer is recharged with surface water from the Santa Clara River through spreading basins. Typical source water characteristics and treated water objectives are presented in Table 2-1. Pretreatment of the source water is required to remove relatively large particulate matter ahead of the RO and NF membranes and to adjust the water chemistry to minimize and control chemical scaling of the membranes.

The source water is drinking water quality (except for the undesirable TDS) and has very low suspended solids. The average silt density index (SDI), a measure of the water's likelihood to cause particulate fouling of the membrane, is very low at less than 0.5, and the turbidity is typically less than 0.2 NTU. The membranes are protected from damage by relatively large particles by an automatic backwashing, bag filtration system with a 20-μm nominal removal. The automatic backwashing filter system was selected to permit continuous plant operation and minimize the operating cost, labor, and maintenance time devoted to the pretreatment screening.

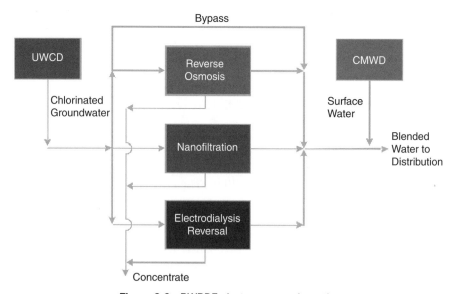

Figure 2-2 BWRDF plant process schematic.

TABLE 2-1 Typical Source Water Characteristics and Treated Water Objectives

Parameter	Average	Maximum	Treated Water Objective
TDS (mg/L)	1000	1300	≤370
Hardness (as CaCO₃) (mg/L)	560	680	≤150
Turbidity (NTU)	0.2	1	≤0.1
SDI	<0.5	1.0	—
Temperature (°C)	18	13–20	—
pH	7.2	7.6	8.2

Note. TDS, total dissolved solids; NTU, nephelometric turbidity unit; SDI, silt density index.

The UWCD had previously added free chlorine to the groundwater prior to delivery to the BWRDF. While the EDR membranes can tolerate low levels of free chlorine, the source water had to be dechlorinated to protect the RO and NF membranes from oxidant damage by free chlorine. Sodium bisulfite, a reducing agent, was initially added to the source water after the pretreatment filter system to reduce the chlorine to chloride. Since 1 June 2000, the UWCD has used chloramines for disinfection residual. The chloramines are passed through the RO and NF membranes to control biofouling. An oxidation-reduction potential (ORP) analyzer is used to monitor the water and provide an alarm if oxidant levels are too high.

Hydrochloric acid is required to minimize mineral scale accumulation on the EDR membranes. Currently the RO and NF membranes do not require acid addition based on the source water pH and TDS levels. Hydrochloric acid can be added ahead of the RO and NF systems should they require pH adjustment in the future. All three membrane systems require a small anti-scalant dose to sequester mineral scale formation. Different antiscalants are fed to each system based on operational experience. Each membrane system has a dedicated acid and antiscalant chemical metering pump to permit different chemical feed rates and to accurately monitor the chemicals used with each of the three membrane systems.

Reverse Osmosis System

The BWRDF RO system is a two-stage process where the concentrated reject stream from the first stage is the feedwater to the second stage (Figure 2-3). The RO system initially had 14 first-stage vessels and 7 second-stage vessels, each with 6 elements per vessel. The RO system was expanded in October 2000 by adding additional membrane elements to the existing skid. The RO system now has 18 first-stage vessels and 10 second-stage vessels. The RO membrane elements are thin-film composite, Filmtec BW40LE-440 elements. The product recovery for the RO system, defined as the product water passing through the RO membranes divided by the source water entering the system, is currently approximately 80%. The pressure required to drive the RO process is about 136 psi. The RO product water has a TDS of about 16 mg/L. This

Figure 2-3 *RO and NF systems at the BWRDF.*

is much lower than the treated water TDS objective of 370 mg/L TDS. To economically produce the desired treated water quality, approximately 0.43 MGD of source water is bypassed around the RO system and blended with the RO product water to produce 1.5 MGD of treated water with the desired TDS.

The TDS concentration in the reject water from the RO, NF, and EDR membrane systems is between 3 and 4 times the TDS concentration in the source water. The reject water from all three membrane systems is currently discharged to the headworks of an adjacent wastewater treatment plant and discharged to the ocean through an existing outfall.

Data collection at the BWRDF is fully automated. The plant Supervisory Control and Data Acquisition (SCADA) system monitors system flow rates, pressures, water quality, chemical use, and power consumption for each membrane system. Plant staff monitor operation and maintenance time and costs for the each of the three systems.

Nanofiltration System

Like the RO system, the NF system is a two-stage process initially with 15 first-stage vessels and 7 second-stage vessels, each with 6 elements per vessel. The NF system now has 18 first-stage vessels and 10 second-stage vessels. The NF membrane elements are thin-film composite, Filmtec NF90-400 elements. The product recovery for the NF system is approximately 80%. The NF pressure required to drive the NF process is about 105 psi. The NF product water has a TDS of about 40 mg/L. This is also lower than the treated water TDS objective of 370 mg/L. As with the RO, source water is bypassed around the NF system and blended with the NF product water to produce 1.3 MGD of treated water with the desired TDS.

Electrodialysis Reversal System

Electrodialysis is an electrically driven membrane process that uses direct-current (dc) voltage potential to drive ions through a semipermeable membrane, reducing the source water TDS. The EDR membrane stack consists of alternating cation and anion transfer membranes. As the source water flows between the cation and anion membranes, the dc voltage potential induces the cations to migrate toward the cathode through the cation membranes, and the anions to migrate toward the anode through the anion membranes. The cations and anions accumulate in the reject water side of the membranes and the low TDS product water is produced. The EDR system periodically reverses the electric field polarity to flush scale-forming ions off the membrane to minimize membrane cleaning.

The EDR system product water does not pass through the desalting membrane as in an RO or NF system. This reduces the potential for particulate fouling on the EDR membrane surface. However, for desalting applications that also require treatment to meet the Surface Water Treatment Rule, the EDR system would require an additional filtration process. Because the source water for the BWRDF meets federal and state drinking water requirements, additional filtration of the EDR product water is not required.

The BWRDF EDR membrane system is an Ionics EDR 2020 (Figure 2-4). The system has five parallel trains, each with three Mark IV membrane stacks. Each membrane stack contains 600 cell pairs of ion-exchange membranes and

Figure 2-4 *EDR system at the BWRDF.*

flow spacers. The product recovery for the EDR system is approximately 85% (some source water is added to the reject water loop to reduce TDS concentrations to minimize mineral scale formation). The EDR system, unlike the RO and NF systems, does not need to bypass and blend source water with the product water to produce 1 MGD of treated water. The EDR system is designed to meet the treated water quality criteria based on an optimum combination of voltage field potential and membrane stacks.

Posttreatment

The BWRDF RO and NF system product water is typically between 5.0 and 5.6 pH units. The EDR system does not reduce pH as much as the RO and NF systems. The EDR product water pH is typically much closer to source water pH. Sodium hydroxide (caustic soda) is added to the product water to raise the pH to approximately 8.0–8.2 pH units for corrosion control. Sodium hypochlorite and ammonia are added to provide a chloramine disinfectant residual in the PHWA's distribution system.

TWO YEARS OF OPERATION

Over two years of operation the BWRDF has produced greater than 3 MGD of high-quality drinking water and helped to reduce overdrafting in the Oxnard Plain Groundwater Basin. However, the first year and a half of operation at the BWRDF was characterized by fouling of the RO and NF membranes. The fouling resulted in higher feed pressures required to produce water and more frequent and costly cleanings. The EDR did not have the excessive fouling problems experienced by the RO and NF and as a result had lower operating costs for that period. PHWA staff were able to identify the cause of fouling and took actions to reduce the problem. In the second year, the BWRDF was also expanded and process improvements were made to make the facility more energy efficient.

Membrane Fouling

The RO and NF systems had problems with bio-fouling and particulate fouling. An autopsy of the bio-slime determined it to be sulfate-reducing bacteria that were most likely using the initial antiscalant (AS-120) and sodium bisulfite as a food source. The solution to the fouling problem was to change antiscalants and to permit chloramines to pass through the membranes to control bio-growth. BWRDF staff tried a number of different antiscalants and found a King Lee product that reduced bio-fouling in the RO and NF membranes. The RO and NF membranes also experienced some particulate fouling. Particle counting and bench scale testing determined that the source water contains 0.25- to 0.5-μm colloidal particles. The bio-fouling was creating a "sticky" surface that was capturing the colloidal particles and leading to clogging in the lead membrane elements. The new antiscalant and switching of membrane elements has minimized this problem. When there is excessive pressure drop across the lead elements, the first membrane element is moved to the last position in the second stage, and the last element in the second stage is placed in the lead position. This procedure tends to clean the fouled elements and permits continued operation at lower feed pressures.

Chloramination ahead of the membranes has had the largest impact in reducing fouling. Figure 2-5 shows the first-stage differential pressure (DP) for the RO and NF membranes for the period from 11 January 2000 to 30 November 2000.

Before chloramination ahead of the membranes, the first-stage DP was 35–40 psi and rose rapidly, requiring frequent cleanings. The RO system was cleaned four times between January and June 2000. In June 2000, the UWCD started to use chloramines as their disinfection residual and the staff at the BWRDF stopped adding sodium bisulfite to the source water. Permitting the chloramines residual to pass through the RO and NF membranes has significantly reduced the cleaning frequency for the membranes and has improved the performance of the system. The RO and NF systems have not required cleaning since June 2000. The first-stage DP has dropped to 20–25 psi and the feed pressure to the RO and NF membranes has dropped by about 30 psi.

NF Membrane Delamination

The NF system was experiencing greater feed pressure increases than was the RO system and was not responding as well to the clean in place improvements and cleaning strategies described above. An NF membrane element was examined and was found to have delamination at the leading outside edge of the element. The membrane manufacturer has replaced the NF membrane elements under warranty.

BWRDF Improvements and Expansion

In the fall of 2000, the RO and NF system capacities were increased by adding membrane elements to the existing skids. Forty-two membrane elements were added to the RO system, increasing its capacity from 1 to 1.5 MGD (product plus blended water). Thirty-six membrane elements were added to the NF system, increasing its capacity from 1 to 1.3 MGD (product plus blended water). Variable frequency drives (VFDs) were added to the RO and NF feed pumps to improve the energy efficiency of the systems, and the recovery for the RO and NF systems was increased from 75 to 80%. The improvements have increased the plant capacity by approximately 30% while reducing power costs by approximately 15%.

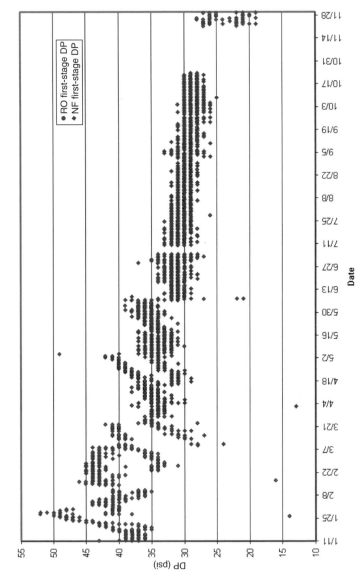

Figure 2-5 RO and NF DP versus time—created at 10:14 on 30 November.

COMPARISON OF MEMBRANE SYSTEM PERFORMANCE

The three membrane systems' operating and maintenance costs and characteristics were monitored during the first two years of operation, between January 1999 and December 2000. The plant SCADA system permits operators to automatically track the usage rate of pretreatment and posttreatment chemicals, power consumption, and system water production for each membrane system. BWRDF staff also maintained records on general labor and maintenance costs, cleaning chemical costs, and system downtimes.

The change to chloramines in the source water in June 2000 resulted in marked improvements in the performance of the RO and NF systems. Furthermore, the capacity of the RO and NF systems was increased and VFDs were added in the fall of 2000. The data presented below are representative of the membrane system performance following the improvements/expansion to the BWRDF. In some cases where marked changes occurred due to the changes, the data are grouped into two categories: (1) data from January 1999 to June 2000, and (2) data following plant expansion and improvements after June 2000.

Water Quality and Production Comparison

During the first two years of operation, the average source water TDS was 1000 mg/L. The BWRDF is designed and operated to produce treated (blended) water with an average TDS of 370 mg/L. This TDS objective matches the typical TDS level in the treated surface water delivered by the Calleguas Municipal Water District. Table 2-2 shows the average source and final blended water quality and compares the product water characteristics from the three membrane systems.

While the ion rejection of the RO membrane is greater than that of the NF membrane, the two systems produced very similar product waters. In terms

TABLE 2-2 Comparison of Membrane System Product Water Quality

Parameter	Source Water	RO Product	NF Product	EDR Product	Blended Water
TDS (mg/L)	1000	16	40	370	370
Conductivity (μS/cm)	1390	23	58	595	595
Hardness (mg/L as $CaCO_3$)	560	8	16	114	179
Alkalinity (mg/L as $CaCO_3$)	220	5	8	148	123
pH	7.2	5.0	5.6	7.2	8.2
Calcium (mg/L)	140	ND	ND	27	46
Magnesium (mg/L)	52	ND	ND	11	16
Sodium (mg/L)	94	3	4	67	53
Chloride (mg/L)	56	2	12	29	25
Sulfate (mg/L)	480	2	2	69	138
Nitrate (mg/L)	20	2	4	8	8

of water quality, there appears to be little distinction so far between the low-pressure brackish water RO membrane and the NF membrane. Table 2-3 presents a comparison of the average membrane system water production before and after the facility improvements.

O&M Cost Comparison

Table 2-4 presents a breakdown and comparison of the O&M costs for each membrane system for performance after the plant improvements/expansion. The costs are presented in dollars per 1000 gallons of blended water (product water plus associated bypass water) produced by each system. The O&M costs in Table 2-4 do not include the cost of distribution system pumping or the cost to purchase source water from UWCD. The costs are described in more detail below.

Labor and power constitute the largest percentages of the BWRDF O&M cost at 56 and 19%, respectively. In the first year of operation, the EDR had the lowest operating cost. However, after the bio-fouling issues with the RO and NF system were resolved and the energy efficiency was improved by adding VFDs, the RO and NF currently have the lowest operating costs at $0.31/kgal.

Labor Cost

Total plant labor cost for the year 2000 was $185,000. This includes two full-time operators and some additional part-time maintenance personnel. The labor efforts and costs for the facility were divided approximately equally between the three technologies. The major labor requirements for the RO and NF systems consisted of daily manual SDIs and periodic CIPs. The labor

TABLE 2-3   Average Membrane System Production

System	Feed (gpm)	Product (gpm)	Reject (gpm)	System Recovery (%)	Bypass (gpm)	Blended Water (gpm)	Overall Recovery[a] (%)
			Before Plant Improvements/Expansion				
RO	673	506	167	75	206	712	81%
NF	712	534	178	75	188	722	80%
EDR	830	690	140	83	0	690	83%
			After Plant Improvements/Expansion				
RO	900	720	180	80	300	1,020	85%
NF	910	728	182	80	200	928	84%
EDR	830	690	140	83	0	690	83%

[a] The overall recovery is the ratio of product plus bypass to feed plus bypass. This gives a fairer comparison of the three processes.

TABLE 2-4 Membrane System O&M Cost Comparison ($/kgal) After Improvements/Expansion

System	Labor	Power	Pretreatment	Posttreatment	CIP	Total
RO	0.179	0.046	0.034	0.05	0.003	0.31
NF	0.179	0.041	0.037	0.05	0.003	0.31
EDR	0.179	0.090	0.031	0.04	0.003	0.34
Overall	0.179	0.059	0.034	0.05	0.003	0.32

required for the EDR system includes wash-downs and stack probing every two weeks and periodic CIPs. The staff at the PHWA have recently instituted a computerized maintenance tracking system to better define the labor required for each of the three systems.

Power Cost

Average power consumption for the three systems is shown in Table 2-5. Controlling the bio-fouling and the VFD pumps on the RO and NF systems has reduced power costs considerably. The current local electrical power cost is approximately $0.07/kWh.

Pretreatment Chemical Cost

Following the plant improvements/expansion, the only pretreatment for the RO and NF systems is antiscalant (KingLee 0100) (Table 2-6). The pretreatment for the EDR system includes hydrochloric acid and antiscalant (Argo AS120). A small amount of excess ammonia is also added to the source water to ensure the RO and NF membranes are not exposed to free chlorine.

Posttreatment Chemical Cost

The average daily posttreatment chemical costs were as follows: $78 for the RO system; $75 for the NF system; and $45 for the EDR system. The cost for caustic soda was divided between the three technologies in relation to the

TABLE 2-5 Average Power Consumption

	Power Use (kWh/kgal)	
Membrane Process	Before Control of Bio-fouling and Improvements/Expansion	After Control of Bio-fouling and Improvements/Expansion
RO	1.64	0.66
NF	1.40	0.58
EDR	1.29	1.29

TABLE 2-6 Pretreatment Chemical Cost

Membrane Process	Chemical	Chemical Usage (gpd)	Total Pretreatment Chemical Cost After Improvements/Expansion ($/kgal)
RO	KingLee 0100	3.7	0.034
NF	KingLee 0100	3.6	0.037
EDR	Argo AS120	0.24	0.031
EDR	HCL	25	

product water pH. The costs for sodium hypochlorite and ammonia are minimal and divided evenly between the three technologies.

Cleaning Cost

The costs of CIP chemicals from January to December 1999 were $841 for the RO system, $12,635 for the NF system, and $1302 for EDR system. The NF system was cleaned with a proprietary (and very expensive) cleaner in an attempt to address significant fouling issues. Defective membranes also exacerbated the NF system fouling. The costs of CIP chemicals from January to June 2000 were $8243 for the RO system, $3323 for the NF system, and $1622 for EDR system. The fouling problems that caused these high cleaning costs have been controlled by the addition of chloramines ahead of the membranes. Since June 2000, the RO and NF have not required cleaning. Based on the performance from June 2000 to December 2000, the RO and NF are expected to be cleaned twice per year and the EDR is expected to be cleaned quarterly. The cost for a quarterly EDR CIP is $515. The cost for an RO and NF CIP depends on the amount and type of proprietary chemicals used. The anticipated costs for CIP of the membrane systems going forward is approximately $2000 per year per system.

SUMMARY

Over the two years of operation the BWRDF produced greater than 3 MGD of high-quality drinking water. In the first year, the facility experience problems with bio-fouling in the RO and NF membranes, and the EDR system had the lowest O&M cost at $0.34/kgal. In the second year operation, changes were made to resolve the biofouling and improve the efficiency of the RO and NF systems and now the RO and NF system have the lowest O&M cost at $0.31/kgal.

Operating challenges during the first two years included fouling of the membranes and delamination. The BWRDF staff has addressed these issues and is continuing to collect data on the long-term operating costs and performance of the membrane systems. The authors intend to update this paper as more data are gathered and operating experience is accumulated.

REFERENCES

1. Thompson, Craig M., Todd Reynolds, and Bill Boegli. New 3-MGD water production & research facility to compare performance & cost for brackish water treatment. *The International Desalination and Water Reuse Quarterly*, 7/2:35–42 (1997).
2. Passanisi, Jim, Janet Persechino, and Todd Reynolds. EDR, NF and RO at a brackish water reclamation facility. *The American Water Works Association Conference Proceedings*, June 2000.
3. Reynolds, Todd K., Frank Leitz, and Jim Passanisi. The startup and first year of operation at the Port Hueneme Brackish Water Reclamation Demonstration Facility. *The American Desalting Association Conference Proceedings*, August 2000.

Desalting a High Total Dissolved Solids Brackish Water for Hatteras Island, North Carolina

Robert W. Oreskovich

Dare County Water Department, Manteo, North Carolina

Ian C. Watson

RosTek Associates, Inc., Tampa, Florida

Dare County, North Carolina, is located on the Mid-Atlantic seaboard of the United States. Given its easterly location, it was an early landfall for the 16th century explorers and is the site of the first English colony in the United States. It is a county that consists primarily of water, and its land area is largely composed of barrier islands. Hatteras Island is the most southerly of the land area and is well known for its lighthouse, one of the largest on the east coast.

Tourism is the basis of the economy, with some commercial fishing. The tourist industry has grown dramatically in the last 15 years, with the bulk of the growth located on the northern beaches. Hatteras Island is home to large tracts of federally owned land, so the growth has been limited to the three villages of Rodanthe-Waves-Salvo, Avon, and the southern end of the island.

Since about 1968, the residents of Avon and the southern end of Hatteras Island have had access to a potable water system. This system was originally owned and operated by the Cape Hatteras Water Association (CHWA), a user-owned cooperative utility. The source of water has been a shallow (40- to 80-ft) well field located in an area known as Buxton Woods.

The current water treatment system has been in operation since February 2000, replacing the former 1.6-MGD system, which dated from 1986. The CHWA tried in the early 1990s to expand the well field to meet future demand, and conducted several studies and pilot tests to bring the water quality into compliance with the then current drinking water standards. However, the well field expansion was challenged on environmental grounds, since the well field is located in a sensitive maritime forest, and also the State determined that the well field was "groundwater under the influence." These two factors severely limited the ability of the CHWA treatment plant to meet or exceed the primary standards for drinking water quality and to supply the quantity of water needed for the continued growth in the service area.

A 1992 study provided an overview of the treatment options available that would permit CHWA a reasonable opportunity of compliance with then current standards, and those upcoming in the future, both known and speculative. The report covered both inadequacies of the current treatment process and process options for future compliance. It did not, however, address the question of raw water resources. In addition to treatment shortcomings, the transmission and distribution system was barely adequate to serve existing customers.

After imposing a moratorium on new connections, the CHWA Board of Directors attempted to find ways of meeting the challenges facing them. One of these efforts resulted in a study by Boyle Engineering Corporation[1] addressing the feasibility of desalting brackish groundwater. As part of this work three test wells were constructed to determine the occurrence, extent, and quality of brackish water sources. The hydrogeologists, Missimer International, found high-yield limestone formations at 200–300 ft, but containing water of highly variable quality. On the basis of the third well, a treatment concept was developed. Although, by necessity, many assumptions had to be made, the treatment process presented was technically viable, with some confirmation of cost required through additional pilot testing. The key to the project was the determination of both the extent and capacity of the limestone formation.

In July 1997, the responsibility for and ownership of the CHWA potable water system passed to the Board of County Commissioners of Dare County. The system became part of the Dare Regional Water System. At this point, engineering contracts were signed with Missimer International for groundwater investigation and well field design; RosTek Associates, Inc. (formerly AEPI/RosTek, Inc.) for design and construction services for a reverse osmosis (RO) water treatment system; and Hobbs, Upchurch and Associates for distribution system upgrades and modernization and for a new water treatment plant facility and structures, including ion exchange treatment and filtration of the existing shallow raw water source. Essentially, the project would provide a completely new water treatment plant for the residents and business and tourist facilities of South Hatteras Island, starting at the Village of Avon, and reaching south and west to the Ocracoke Ferry Terminal.

The Future Water Supply Study identified a treatment system that would combine the use of some of the existing shallow wells and treatment plant equipment with a totally new brackish water RO plant. The existing well water, highly colored and with variable iron content would be treated for color/total organic carbon reduction by anion exchange, and then oxidation and filtration for iron removal. Pilot testing confirmed both the process parameters and cost for both parts of the treatment process.[2] By significantly reducing the trihalomethane formation potential (THMFP) and virtually eliminating the iron, the natural calcium hardness and the alkalinity of the shallow well water would be used advantageously by blending with RO permeate, which contains virtually no hardness and alkalinity.

As the hydrogeological studies progressed, two things became clear. First, the limestone aquifer was proving to be extremely productive, and second, the quality was very variable from site to site. Solute modeling predicted a steady increase in total dissolved solids (TDS) with time, reaching a limiting value after about 15 years of about 15,000 mg/L TDS! This required that the RO system be designed to operate in the early years at a higher recovery, with the recovery gradually declining as the feedwater TDS increased. To complicate the exercise even more, it was decided to incorporate energy recovery devices on each train and to add membrane area as the TDS increased to maintain a relatively constant feedwater pressure.

As the pieces of the puzzle started to fall into place, the key decisions as to capacity, number of trains, membrane array, maximum salt passage, and so on were made. Earlier estimates and Planning Department studies showed that a build-out production capacity of 3.0 MGD would be required to meet the needs of the permanent residents, summer visitors, and fall fishermen. The shallow well field water production (to be used for blending with RO permeate after treatment) was limited to 1.0 MGD, as a result of several hydrogeologic studies prepared by CHWA during the Buxton Woods hearings. In fact, blending studies showed that a ratio of 2.33:1, permeate to shallow, would provide a well-buffered blend. This led to the selection of 2.1 MGD of RO permeate capacity, blended with 0.9 MGD of treated shallow well water. The RO was designed in three trains, with two installed initially.

As a result of these considerations, a decision was made by the County to limit shallow well production to 0.9 MGD maximum day and to provide for 2.1 MGD of RO capacity. It was further decided to install this capacity incrementally, with initial installed capacities of 0.6 and 1.4 MGD, respectively.

Selection of recovery, membrane type, and operating characteristics were very much dictated by the circumstances surrounding this project. Because the blend water from the shallow well field will have a TDS lower than 500 mg/L, with good calcium hardness and alkalinity, it was decide to use a standard pressure high rejection brackish water membrane for the RO portion of the treatment process. The projected permeate quality at startup was relatively low, and as the feed water salinity increased with time, the plant recovery would be lowered to maintain a relatively constant concentrate quality. This was fairly critical, since the concentrate discharge location is on the sound or north side of Hatteras Island at that point, and the sound water TDS averaged about 26,000 mg/L. This became a limiting factor in the selection of the initial recovery, which turned out to be 70%. Based on the projections of future well water quality, at the limiting TDS of about14,700 mg/L, recovery would have turned down to about 50%. Clearly this presented a compelling argument for the inclusion of energy recovery devices, which were installed in the plant.

In the design phase the decision was made to specify an RO feed pump that could accommodate both sets of conditions, initially 70% recovery, declining to 50% recovery. Further, the feed pressure to the membrane system was to remain fairly constant at around 400–425 psig. This was accomplished in the final plant configuration, with a relatively efficient vertical turbine pump, equipped with a 250-hp high-efficiency electric motor.

In selecting an energy recovery turbine for this application, the primary criterion was the ability of the device to accommodate the future condition with major rework. The Pump Engineering Turbo was selected, with the obvious compromise that the initial operation would be less efficient than if the design point for the plant was fixed. However, using the projected rate of increase in feedwater TDS, and the corresponding change in operating conditions, the payback at the current power rate of $0.12/kWh turned out to be about 4.5 years. If the feedwater TDS were to increase more rapidly (this in fact has happened!), then the payback period would be shorter.

Because of the changing recovery, the decision was made to utilize a single-stage RO array, with seven elements per vessel. For the initial 70% recovery, the β-factor was less than optimal, but became acceptable at about 68% recovery. Four hundred square foot elements were used, with 16 vessels initially, increasing to 24 at final configuration. The average flux initially is 15.6 gfd, declining to a possible 10.4 gfd, if all 24 vessels are installed. The membrane selection was left up to the system supplier, with Hydranautics CPA-3 elements being supplied.

On the shallow well water side of the plant, the general contractor was given the choice of either supplying new vessels for the ANIX system and filters or using the vessels from the existing plant, removing them off-site for refurbishing, and reinstalling in the new building. The latter option was taken, which meant that for a period of time, a small amount of brackish well water had to be blended with the RO permeate to reintroduce some hardness and alkalinity. This blend could only be about 20 gpm/RO train, because of chloride limitations.

Because of the high TDS feedwater, a flushing system using permeate was included in the specification to evacuate the high chloride water from the RO systems at shutdown. Included in this system is an ultraviolet sterilizer (Figure 3-1). After some discussion, this device was included to provide a measure of protection to the membranes in case the stored permeate became biologically active. Hopefully, this system will turn out to be an unused insurance policy. The stored permeate, about 10,000 gallons, is also used as a chlorine-free source of water to backwash the ANIX beds and as makeup for the ANIX brining system (Figure 3-2).

The RO permeate is blended with the treated shallow well water prior to the addition of posttreatment chemicals as the water leaves the plant building on its way to the 3.0-MG ground storage tank. Posttreatment systems include caustic soda, hydrofluosilicic acid addition, disinfection with sodium hypochlorite, and the addition of a corrosion inhibitor. Since no acid is used in the RO pretreatment step (scale control depending on a high-performance inhibitor), CO_2 stripping is not required, so there is no clear well or transfer pumping system. In fact, none of Dare County's RO plants use CO_2 stripping.

The plant was installed during 1999 and started up with the usual euphoria and despair! A series of events tempered the rookie operating staff early on, including immediate leaks in pressure- and leak-tested piping systems, which always turned out to be acid, fluoride, or some other unfriendly material; malfunctioning instruments; communications problems with well control systems; and a blown 16-inch plastic pipe spool where the feedwater enters the building. When these normal but very tiresome problems where taken care of, and after several false starts, the RO units ran and made water. Two things

Figure 3-1 *Permeate flush tanks, pumps, and UV sterilizer.*

Figure 3-2   ANIX vessels on right, sand filters on left.

were immediately noticeable: (1) the permeate quality was not as expected, and (2) the energy recovery turbine (ERT) was not providing the boost or the recovery promised. The first problem was solved by replacing the last two elements in each vessel with elements having higher chloride rejection than the CPA-3, but requiring a higher net driving pressure to produce the flow. This in turn resulted in an increase in feed pressure, which actually mitigated the ERT performance shortfall somewhat. The ERT is located on the RO feed pump discharge (Figure 3-3), and the increased boost resulted in only a small increase in pump discharge pressure.

Four wells supply the RO plant, each having different quality and pumping rates. Therefore, the combined feedwater entering the plant has variable TDS,

Figure 3-3   Energy recovery turbine in foreground, RO feed pump in background.

depending on the wells in service at the time. Table 3-1 show the various qualities available at start-up, pumping rates, and available combinations for a two-train operation.

During well flushing and disinfection, the well water conductivity started to rise in wells RO-1 RO-2. During this time SDI tests were conducted periodically on each of the wells and on the combined water entering the plant. SDIs for RO-2 and RO-4 were consistently high, and soon after startup both were inspected with a down hole camera. It was clear from the video that the limestone in these wells was soft and interspersed with shell fragments and sand pockets. Some cavities had already developed. After consultation with the hydrogeologist, it was determined that the stress induced on the limestone surface when the well started was causing sloughing of the material, and this was clearly seen in the video. RO-4 eventually cleaned up to the point where the SDI was fairly consistently below 4, but RO-1 was taken out of service shortly after plant startup in February 2000. Well control valves were installed at the wellheads, and the pumping rates were adjusted downward for wells 2 and 4. This appears to have mitigated the problem. SDI values downstream of the 5 μ cartridge filters have been consistently less than 2, but cartridge filter life has apparently not suffered.

The RO plant operation has been consistent since the startup, with feedwater conductivity ranging from 13,000 to 21,000 μs. Permeate quality has consistently been below 500 μs. However, because of the rise in feedwater TDS, which now appears to stable, additional membrane area must be added to the plant this year, about three years ahead of schedule, and the recovery adjusted down to about 67%.

On the other side of the aisle, the shallow well system went into operation in May, and after the initial teething problems, performed well. Color has been consistently reduced to below 5, from an average influent value of 30–80, depending on the wells in operation. Iron reduction has also been as expected with iron in the effluent from the filters averaging 0.2 mg/L. Influent iron typically is around 1.0 mg/L. As the operators gain experience with regeneration cycles and start to optimize salt-loading rates, the operation will become more consistent. Unfortunately, the ANIX system suffered from lateral failure after about four months of operation, and a significant amount of

TABLE 3-1 Start-up Conditions

Well No.	TDS (mg/L) Initial	Current	Pumping Rate (gpm)	Well Combinations	Feed TDS mg/L	Feed Conductivity μs
RO-1	6490	11,200	820	RO-2,3	7,574	13,405
RO-2	8687	12,760	900	RO-2,4	6,605	11,690
RO-3	5527	7,860	500	RO-1,4	5,311	9,483
RO-4	3894	6,840	700			

TABLE 3-2 Finished Water Quality, Before pH Adjustment

Ion	IX Product	Permeate	Blend
Ca	92.00	0.90	28.23
Mg	8.80	2.00	4.04
Na	17.00	72.20	55.64
K	3.00	3.20	3.14
HCO_3	287.00	6.00	90.30
SO_4	10.00	2.70	4.89
Cl	70.00	115.20	101.64
TDS	344.00	202.20	244.74
Ca H	229.00	3.39	71.07
Total Alk	235.00	4.92	73.94
pH	7.60	5.74	6.97
CO_2	11.00	17.70	15.69
LSI	0.40	−4.10	−1.15

Note. TDS, total dissolved solids; LSI, Langelier saturation index.

resin was carried into the filters. The repairs are being made and the system will be put back into operation.

As yet, it is not possible to accurately compile cost data for the operation, but the estimated cost for operations, excluding debt service, was $1.25/kgal of blended water. The CHWA plant operating cost was between $5 and $6/kgal, a good part of which was the cost of hauling chemicals to the end of Hatteras Island and hauling the water treatment plant sludge over 90 miles to a landfill! The total project was constructed for around $9 million, including wells, well housings, well site acquisition, outside piping, and the plant build-

Figure 3-4 *The finished facility.*

ing and process equipment. A 3-MG ground storage tank was also included in the work, together with a separate high service pumping system.

This was not an easy project to design or build. The location is remote and the construction was interrupted in 1999 by one of the busiest hurricane seasons in many years. The hydrogeology of the limestone formation was unknown at the start, and the learning curve for predicting quality has been steep. However, the design goals of the RO portion of the plant have been met, and it is fully expected that the shallow well treatment system will be reliable and effective. The finished water quality (Table 3-2) is excellent, a vast improvement over that which was being delivered to the customers from the old treatment plant. Figure 3-4 depicts the finished facility.

REFERENCES

1. *Future Water Supply Study.* Santa Rosa, CA: Boyle Engineering Corporation, September 1995.
2. Oreskovich, Robert W., et al. Planning a new water supply for Cape Hatteras, NC. ADA Biennial Conference, Williamsburg, Virginia, August 1998.

Case Study 4

Desalination for Texas Water Supply

Mark Graves
HDR Engineering, Inc., Austin, Texas

Bryan Black
HDR Engineering, Inc., Portland, Oregon

James Jensen
Parsons Brinckerhoff Water, Tampa, Florida

James Dodson
Nueces River Authority, Corpus Christi, Texas

Gary Guy
San Antonio Water System, San Antonio, Texas

Water desalination is an increasingly attractive option to produce potable water for the growing Texas population. Technological advances in desalination, shifting market conditions, and increasingly stringent drinking water treatment regulations are making desalination more attractive relative to conventional drinking water treatment. Desalination of seawater in Texas has the potential to expand the resources available for producing potable water. It is increasingly difficult to develop freshwater storage projects, particularly in-channel reservoirs. Additionally, the value of interbasin water rights transfers has been diminished. Population growth continues even in areas vulnerable to drought where freshwater is limited. These factors are driving water utilities and industry to consider desalinating seawater in Texas.

The Tampa Regional Water Supply project for a 25-MGD seawater reverse osmosis (RO) system received proposals with water costs 2–3 times lower than those previously observed for other large-scale seawater desalination facilities. These low costs resulted not only from technological improvements, but also from siting and macroeconomic factors. Information from the Tampa project was gathered and reviewed for this report to determine the factors leading to this major advance in seawater desalination and their potential application along the Texas coast. These factors were incorporated into a cost-estimating model and a general framework was developed for making siting decisions for seawater desalination on the Texas coast. Potential environmental impacts and permitting issues for a desalination facility were evaluated and included for consideration of project feasibility.

Research was also conducted to review membrane technologies and costs in general for desalination of both brackish waters and seawater.[1] Reverse osmosis and electrodialysis reversal (EDR) are the primary membrane treatment processes currently implemented to remove dissolved salts from water. This study focuses on desalination of high-salinity waters either from the ocean or from mixed bay systems. Therefore, findings and information are based on the use of reverse osmosis because EDR is generally not considered for desalination of waters with greater than 3000 mg/L total dissolved solids (TDS).

REVERSE OSMOSIS DESALINATION BASIC CONCEPTS

State and federal regulatory agencies require that drinking water meet primary drinking water standards. The voluntary secondary drinking water standards limit constituents in water that affect the aesthetic quality of drinking water, such as taste, odor, color, mineral content, and appearance, that may deter the public acceptance of drinking water. Membrane desalination technologies can demineralize water so that secondary standards are met, producing water with a pleasing aesthetic quality. Reverse osmosis membrane filtration produces superior water that can meet even the most stringent primary drinking water regulations.

Desalination provides economic benefits and enables wastewater reuse. Due to perceived health impacts or taste preferences, customers may treat mineralized water with home treatment units or use bottled drinking water. Industry may be forced to install point of entry treatment for pure process water. Providing centralized desalination treatment eliminates the need for site-specific treatment. A mineralized water supply produces a mineralized wastewater, restricting the reuse of wastewater for agricultural irrigation. Therefore, desalinating water using membranes in a central facility can reduce costs to the homeowner or industry and provide wastewater effluent that is more suitable for reuse.

Desalination processes are generally more expensive than conventional water treatment but the costs are decreasing due to a more competitive market and technological innovation. Desalination has two principle steps: water-concentrate separation and concentrate disposal. Figure 4-1 shows a schematic of membrane desalination systems.

TAMPA BAY SEAWATER DESALINATION PROJECT

Two recent contracts highlight the potential for low-cost seawater desalination. In July 1999, Tampa Bay Water entered into a water purchase agreement with the development team S&W Water, LLC to fund, design, build, operate, and, at some point, transfer a seawater desalination plant. The plant is to have an installed capacity of 29 million gallons per day (MGD), producing an average of 25 MGD of potable water at an average cost over 30 years in present day dollars of $2.08 per 1000 gallons. This cost is two to three times lower than costs previously observed for large-scale seawater desalination facilities. Also, in late 1999, the Water and Sewerage Authority of Trinidad and Tobago contracted with an Ionics, Inc. joint venture to design, build, and operate a seawater desalination plant. This plant is to produce 28.8 MGD of potable water at an average cost over 23 years of $2.67 per 1000 gallons.[2] The low-cost factors for the Tampa Bay Water project are evaluated here to provide background for application of these factors in Texas.

Design/Build/Operate

The design/build/operate project delivery option offers many advantages for seawater desalination contracts. Seawater desalination facilities must be customized to treat source waters with variable water qualities to deliver product water that meets client/customer specifications. In most cases process parameters cannot be determined without extensive pilot testing and then process parameters may need to be modified once full-scale operation begins. These types of projects lend themselves to the performance-based contract process, where the water quality, quantity, delivery schedule, and so on are specified but the plant design is left to the developer. Performance-based specifications allow developers to propose the best and most cost-effective technology that

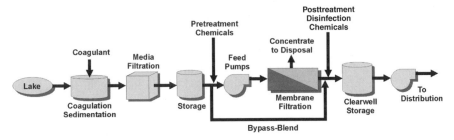

Figure 4-1 *Typical surface water desalination schematic.*

they are familiar with. It also allows for the project to take advantage of innovations in desalination technology, which also generally lowers the cost. Design/build/operate also transfers more of the project risk to the developer in that the developer specifies the plant design and yet must meet the performance specifications.

Power Plant Colocation

The Tampa Bay Water desalination plant will avoid substantial capital costs by sharing the intake and outfall canals with the Tampa Electric Company power station. The feedwater for the desalination plant will flow through the trash grates and screens of the power plant. Underwater construction is avoided in that the intake and discharge pipeline from the desalination plant tie on land into the power plant cooling water discharge pipeline. The elevated temperature of the discharged cooling water (approximately 15°F above ambient Bay water temperature) will increase the amount of product water produced by the membranes in the desalination plant.

The power plant cooling water flow is approximately 1350 MGD, providing dilution for the 16.7-MGD concentrate discharge flow. Due to the high rate of dilution the salinity in the power plant effluent is expected to rise by less than 2%. Without this large cooling water flow it may not be possible to discharge the concentrate into the bay without additional mixing facilities. It is estimated that $15 to $130 million in capital cost avoidance and considerable O&M cost saving was realized due to colocating the desalination plant with the power plant. Table 4-1 summarizes approximate cost savings for colocation with the power plant.

Source Water Quality

Favorable water quality (lower TDS) of the raw water will contribute to decreased operating costs (principally, lower electric power requirements). Analysis indicated that TDS at the intake ranged from 10,000 to 33,000 mg/L, with an average annual salinity of about 26,000 mg/L. This is considerably lower than the typical open ocean TDS of approximately 35,000 mg/L. However, because of the fluctuating TDS concentration, variable frequency drives are required for the high-pressure pumps at an additional capital cost.

The surface water source for the desalination plant has a relatively high fouling potential due to biological activity in the bay and erosion runoff (sediment) into the bay. However, the Big Bend intake canal is approximately 3460 ft long, 200 ft wide, and 20 ft deep, with a water flow velocity of about 0.5 fps. Therefore, even with high suspended solids loading in the bay, the intake channel will act as a settling basin to allow the majority of sand and silt to settle out. The algae and other biological matter have significant fouling potential requiring a high-capacity pretreatment system to protect the reverse osmosis membranes. A budget of approximately $13,318,000 was set aside for the feedwater pretreatment system for the desalination plant.

TABLE 4-1 Tampa Bay Power Plant Colocation Cost Savings

	Low Estimate ($)			High Estimate ($)		
	Capital Cost	O&M Cost	Cost per 1000 Gallons	Capital Cost	O&M Cost	Cost per 1000 Gallons
Intake canal	5,000,000	1,000,000	0.15	40,000,000	2,000,000	0.54
Outfall canal	5,000,000	1,000,000	0.15	40,000,000	2,000,000	0.54
Trash gates and screens	300,000	30,000	0.01	500,000	300,000	0.04
Elevated temperature[a]	4,000,000	250,000	0.06	7,563,492	334,106	0.10
Data and modeling for permits	1,000,000	100,000	0.02	2,000,000	100,000	0.03
Ongoing monitoring	0	100,000	0.01	0	300,000	0.03
Total	15,300,000	2,480,000	$0.39	130,063,492	5,034,106	$1.59

[a] Water flux increases by 2%/°F temperature increase. Cost savings for temperature increase based on 15°F increase resulting in flux rate increasing from 6.46 to 8.4 gal/sfd for 25-MGD product water flow rate with 168 × 8 element array (1344 elements). The average Bay temperature is 77°F and the average boiler condenser discharge used for feedwater is 92°F.

Assumptions: Interest rate = 6.0%, financing period = 30 years, average product flow = 25 MGD.

Environmental Conditions, Permits, and Mitigation Requirements

Extensive agency review is anticipated due to a lack of precedence in permitting in the United States a desalination facility of the size and configuration of the Tampa Bay project. However, the effort required by the developer to fully meet all environmental data acquisition and modeling requirements will be diminished at the selected site due to previous permits and studies required for the existing power plant. Additional savings for the developer will be realized due to studies conducted in the Bay for other purposes and studies conducted on behalf of Tampa Bay Water during the desalination proposal selection process. A budget of $1,300,000 has been established by the developer for obtaining the required permits for the desalination plant and pipeline.

Another advantage of the Tampa Bay location is the large amount of flushing that occurs in the Lower Hillsborough Bay where the Big Bend Power Station cooling water discharges. A study by the U.S. Geological Survey concluded that with each tide reversal, more than 25 times as much water enters or leaves Hillsborough Bay than is circulated through the power station.[3] The overall residence time for Tampa Bay is approximately 145 days.[4] However, the Big Bend Power Station discharges to the lower portion of Tampa Bay near the interface with the open Gulf, and therefore the overall residence time for all of Tampa Bay may not be representative of flushing that occurs near the Big Bend Power Station. Without adequate flushing it would not be possible to discharge the concentrate into the bay due to the risk of salinity buildup causing ecological damage.

SITING ISSUES ASSESSMENT

Siting issues for desalination facilities on the Texas coast were evaluated by four interdependent methods. First, cost models were developed to quantify the effects of major source water, siting, and macroeconomic parameters on product water costs. Second, Geographic Information System (GIS) figures and data tables were used to summarize environmental features and siting conditions along the Texas coast. Third, regulatory and permitting issues relevant to siting a seawater desalination facility along the Texas coast were discussed. Finally, all of the information gathered on cost impacts, siting conditions, and regulatory considerations was used to assess the costs and viability of siting a seawater desalination facility at example sites on the Texas coast.

Capital and Operation and Maintenance Cost Models

In addition to example costs obtained from the Tampa Bay Water project and other desalination projects, two separately developed cost models were used to analyze desalination costs. The use of cost models allows the flexibility to test cost sensitivities for varying process parameters and site specific conditions. The two models were used in conjunction to estimate different portions of the cost analysis and also as a check against each other. Additional costs not covered by either of the cost models are estimated using a combination of engineering calculations, historical costs, and information from manufacturers.

The American Desalting Association and U.S. Bureau of Reclamation distribute a model developed in Microsoft Excel that is titled Water Treatment Estimation Routine (WaTER). WaTER is based primarily on the Environmental Protection Agency (EPA) report, "Estimating Water Treatment Costs," Vol. 2, "Cost Curves Applicable to 200-MGD Treatment Plants" (EPA-600/2-79-1626, August 1979). This is a detailed cost model that can be used to calculate desalination system costs using several different treatment processes, including reverse osmosis, nanofiltration, ion exchange, and electrodialysis. Included are costs for other pretreatment and posttreatment processes relevant to desalination, such as gravity filtration and lime feed. Model input is specific water quality parameters, such as TDS concentration, pH, and alkalinity, along with general input such as flow and recovery rate. From this input the model calculates the cost of a treatment process for particular source waters. The model does not include means to estimate costs for energy recovery turbines, source water intake, concentrate disposal, or delivery to the point of distribution.

The second model used is based on a document currently being developed for EPA entitled "Manual of Cost Estimates for Selected Water Treatment Technologies." Cost information from the EPA document was developed into a model based on standard desalination costs using reverse osmosis. This model includes standard reverse osmosis water production costs for feedwater pumping, pretreatment (acid and antiscalant addition and cartridge filters), reverse osmosis membranes and process system, and membrane cleaning system.

GIS Mapping

GIS coverages available for download on the Internet were used to evaluate and present environmental and geographic information relevant for siting a desalination facility along the Texas coast. Several government agencies supply GIS information on their web sites for general use. Some of the agency web sites where information was obtained include those of the Texas General Land Office, Texas Natural Resource Information System, and Texas Parks and Wildlife Department.

Regulatory Considerations

To better understand the regulatory considerations for siting a seawater desalination facility on the Texas coast, sources of information on desalination regulations and previous projects or studies were reviewed.

DESALINATION COST IMPACTS IDENTIFIED

The cost impacts of different siting parameters were estimated using developed cost models, engineering calculations, and example projects. Both initial capital expenditures and annual operations and maintenance costs are included in the cost-impact analyses. Some siting parameters have a general impact on the entire desalination process and are quantified by estimating the impact on water production costs. Other siting parameters impact only a particular portion of the desalination process and are quantified by their impact on those individual components of the water system. The term "water production costs" is used to refer to the core desalination process without the other ancillary components of a complete water supply system. Water production costs include standard water treatment components common to all seawater reverse osmosis systems, such as feedwater pumps with energy recovery turbines, standard pretreatment (acid and antiscalant addition and cartridge filters), RO membranes and process system, and membrane cleaning system. Since the cost models do not include energy recovery turbines, these were estimated using engineering calculations and historical costs. Water production costs do not include other costs that are more site-specific, such as costs for source water intake, additional pretreatment (e.g., chlorination/dechlorination or media filtration), posttreatment, concentrate disposal, or delivery to the point of distribution. These excluded items may have significant cost implications and are considered separately.

Parameters of the Tampa Bay Water desalination project were used as the base assumptions in most of the estimated example costs. The base assumptions used in the cost estimates are given in Table 4-2. These are the base assumptions used for all the variables in the estimates except where noted in the individual cost-impact estimates. The cost impacts of a few of the major siting parameters are included in the following portions of this section.

Source Water Salinity

Source water salinity affects almost every aspect of the RO process. Required driving pressure across the membrane is dictated by the osmotic pressure caused by the difference in salinity concentrations between the feed and product waters. Increased feedwater salinity increases the osmotic pressure, requiring higher driving pressure. Higher operating pressures necessitate the use of stronger membrane pressure vessels and RO elements designed to handle higher operating pressures.

Recovery rate and process configurations are also affected by source water salinity. Higher salinity generally decreases the recovery rate of a single-stage process configuration. Depending on the source water salinity and required product water TDS concentration, different levels of reject staging, product staging, or bypassing/blending staging may be necessary. High TDS source

TABLE 4-2 Base Assumptions for Estimates

Parameter	Assumption	Description
Labor, including benefits	$25 per hour	
Energy cost	$0.04 per kWh	Interruptible power
Interest rate	6%	
Financing period	30 years	
Recovery rate	60%	Percentage of feedwater recovered as product
Flux	8.4 gfd	Rate product water passes through membrane
Pumping head	900 psi	Pressure for seawater
Cleaning frequency	6 months	Membranes cleaned once every 6 months
Membrane life	5 years	Membrane elements replaced every 5 years

water will produce higher TDS reverse osmosis concentrate that may be more difficult to dispose of due to permitting issues.

Water production costs versus feedwater TDS are shown in Figure 4-2. These costs are based on increasing feedwater pressure with increasing TDS concentration. Feedwater pressures vary from 400 to 900 psi as the TDS concentrations increase from 10,000 to 35,000 mg/L, with the pressure increasing by 100 psi for each 5000 mg/L increase in TDS. The costs are based on constant flux rate of 8.4 gfd and recovery rate at 60% regardless of TDS concentration. Curves could be significantly steeper if process configuration

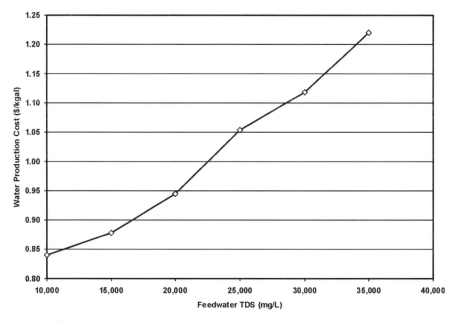

Figure 4-2 Reverse osmosis water production cost versus feedwater TDS.

and/or product water quality requirements cause a decrease in flux rate and/or recovery rate in response to higher TDS concentrations.

Feedwater pump capital costs and energy consumption assume the use of energy-recovery turbines to recover some of the energy in the concentrate. Capital costs of the energy recovery turbines are assumed to be 50% of the feedwater pumps capital cost. It is assumed that 65% of the energy in the concentrate is recovered. Therefore, energy recovered is a function of the recovery rate and feedwater pump energy:

Energy recovered Feedwater pump energy $\times$ (1 $-$ Recovery rate) $\times$ 65%

Figure 4-3 shows a schematic of the energy-recovery turbine system.

Source Water Fouling Potential

Reverse osmosis membrane elements are susceptible to fouling that can decrease the flux rate through the membrane thereby decreasing the treatment capacity per element or requiring higher operating pressures to maintain production. Sources of fouling include suspended solids, organic matter, microbial growth, and inorganic scale deposits.

Source waters with a higher fouling potential can also increase desalination costs by requiring higher levels of pretreatment and/or membrane cleaning. Pretreatment may include chlorination/dechlorination, acid addition, antiscalant, and cartridge filters. Poor source water quality can also require additional pretreatment, such as chemical coagulation, media filtration, and/or ultrafiltration (low-pressure membrane filtration). The required frequency of membrane cleanings may increase with higher fouling potential. Also, some fouling agents are difficult, if not impossible, to remove by current cleaning methods, thereby shortening the effective life of the membranes requiring more frequent membrane replacement.

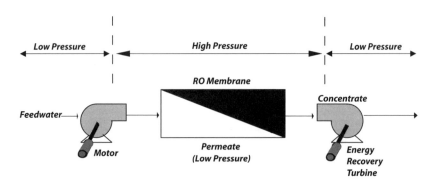

Figure 4-3 *Energy-recovery turbine schematic.*

Feedwater characteristics used to predict fouling potential include pH, alkalinity, temperature, and concentrations of several constituents. The pH affects alkaline scale formation, membrane stability, and salt rejection optimization. Lowering pH by acid addition to about 5.5–6.0 so the Langlier index is negative can reduce the scaling potential due to calcium carbonate. Temperature affects flux rates, membrane life, and scaling. Elevated levels of water constituents, such as strontium, barium, iron, hydrogen sulfide, and silica, can impair performance of RO membranes. The fouling potential of source water can also affect the flux rate achieved across the RO membrane elements. Lower flux rates require more membrane elements or operating at a higher pressure to produce the same quantity of product water.

Concentrate Disposal

One of the most contentious siting factors for a large-scale desalination facility is determining an acceptable location to discharge the concentrate. Potential concentrate disposal methods include discharge to a bay or open ocean, deep well injection, solar ponds, thermal evaporation, and discharge to sewer system. With seawater desalination recovery rates ranging from 40 to 60% there can be a high volume of concentrate generated. Example concentrate production quantities and qualities with varying recovery rates are shown in Table 4-3. For large seawater desalination facilities the only practical option for concentrate disposal may be discharge to a bay or open ocean. Other options may be feasible for smaller plants (less than 5 MGD) where the volume of concentrate is less prohibitive for other disposal options.

A study[5] for the Tampa Bay Water desalination plant indicated that an increase in salinity of less than 6% above baseline in the receiving surface water is most likely not detrimental to native biota. Current EPA regulations allow for an increase of no greater than 10% in background salinity concentration. Additional studies by the Florida Department of Environmental Protection and others have also shown that, with sufficient dilution, desalination concentrate can be discharged to marine waters with negligible impact to the surrounding environs.[6] However, site-specific studies are necessary to characterize existing conditions and to quantify potential impacts to water quality and living resources resulting from a desalination facility at sites along the Texas coast.

TABLE 4-3   Concentrate Production

Recovery Rate	40%	50%	60%	70%
Feedwater flow (MGD)	62.50	50.00	41.67	35.71
Concentrate flow (MGD)	37.50	25.00	16.67	10.71
TDS of concentrate (mg/L)	50,000	60,000	75,000	100,000

Note. Source water TDS = 30,000 mg/L; product water flow = 25 MGD.

Typical concentrate production values are shown in Table 4-3. The volume of concentrate decreases as the recovery rate increases. However, when concentrate volume is reduced, dissolved solids in the concentrate are more highly concentrated. Depending on disposal method and regulatory considerations it may be more or less advantageous to have a greater volume with lower concentration. For highly concentrated discharge, a mixing zone may allow surface discharge of the concentrate. However, disposal of highly concentrated discharge may be limited by bioassay test requirements. Where there are allowances for a mixing zone, the maximum concentration within the mixing zone is dependent on the acute toxicity concentration. The concentrate at higher recoveries may exceed the allowable toxicity concentration.[7]

Concentrate disposal costs can vary widely depending on regulatory requirements and disposal method utilized. Disposing of concentrate through a co-sited outfall, such as the power plant outfall proposed in Tampa Bay, can dramatically decrease concentrate disposal costs. However, concentrate disposal costs can be a large portion of the total desalination cost if more costly options such as offshore discharge are required.

Estimated offshore concentrate disposal costs are shown in Figure 4-4. Costs are based on disposing of 16.7 MGD of concentrate, which is the concentrate from a seawater desalination plant producing 25 MGD of product water with a recovery rate of 60%. The offshore disposal system consists of concentrate pumps, 42-inch pipeline laid on the ocean floor in a 6-ft deep trench and covered, and a diffuser array at the end of the pipeline. Pumps are sized to provide a residual pressure of 100 psi at the end of the pipeline to

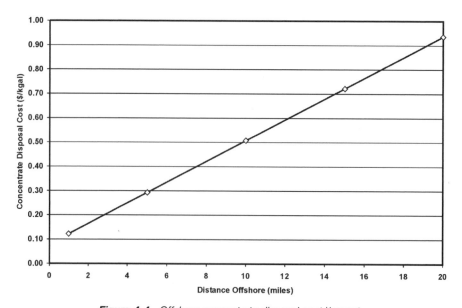

Figure 4-4   Offshore concentrate disposal cost/impact.

allow sufficient concentrate exit velocity from the diffuser nozzles for mixing. Sea grass mitigation costs are included, assuming that 50% of the disposal line will be laid in sea grass areas. Mitigation is assumed to consist of replacing five times the sea grass area disturbed. From previous project experience, mitigation cost is estimated to be $200,000 per acre of sea grass area disturbed. An additional 10% of the construction cost is added to account for potential environmental studies and reports. Costs are shown as dollars per 1000 gallons of product water (25 MGD or 28,000 AFY).

Some of the offshore concentrate disposal cost information was derived from an offshore brine disposal project associated with the storage facility at the Bryan Mound Salt Dome that was part of the Strategic Petroleum Reserve (SPR) Program that started in 1975 and was implemented by the Department of Energy. The Bryan Mound SPR site is located in Brazoria County near Freeport, Texas. The Bryan Mound project consisted of storing petroleum reserves in underground caverns previously filled primarily with salt. The salt from the caverns was leached out with water diverted from the Brazos River. A pipeline and diffuser was built to dispose of the concentrated brine in the open Gulf of Mexico.[8] Construction costs for the 36-inch pipeline and diffuser only with costs updated to March 2000 were approximately $2,500,000 per mile[9] for a construction cost of $31,250,000 for the 12.5-mile pipeline. This cost does not include construction costs for pumping and other miscellaneous costs for the project, such as design and permitting.

Power Cost

Seawater desalination is a power-intensive treatment process, so desalination costs are highly sensitive to the price of power. Power costs are generally about 30% of total seawater desalination costs. Electrical consumption for state-of-the-art RO seawater desalination with energy recovery can range from about 11 to 19 kWh per 1000 gallons of product water. Use of energy-recovery turbines (ERT) can significantly reduce power requirements by recovering up to 85% of the energy remaining in the concentrate. Stone & Webster's Tampa Bay proposal indicates that for their desalination facility the energy-recovery turbines will recover about 26% of the total power used by the feedwater high-pressure pumps (HPRO pumps 13.3 kWh/kgal, ERT −3.5 kWh/kgal). Because the RO process can be easily started and stopped, interruptible power can typically be used provided adequate on-site water storage facilities are provided. The relative impact of power cost on the RO water production cost is shown in Figure 4-5.

All the base assumptions shown in Table 4-2 are used to determine the relative impact of power cost. The feedwater pumps consume the majority of power. Energy required is dependent on several factors, including the salinity and related feedwater pressure and also the recovery rate that affects the amount of feedwater that must be pumped. The impact of recovery rate on the quantity of power required is somewhat mitigated with the use of efficient

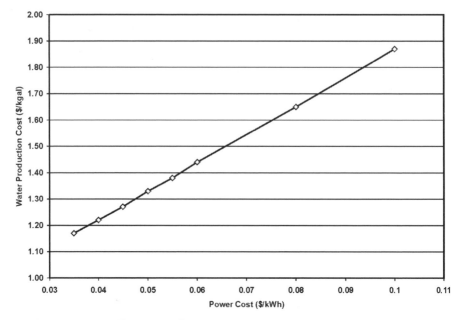

Figure 4-5 *Reverse osmosis power cost impact.*

energy recovery turbines. The costs assume that energy-recovery turbines that recover 65% of the energy in the rejected concentrate are used.

Product Water Flow

The quantity of water to be treated has an impact on total water costs. Significant savings can be realized from efficiencies present in facilities producing larger quantities. Figure 4-6 shows the relative impact of product water flow versus water production cost for flows from 1 to 50 MGD. Energy-recovery turbines are included for product water flows of 5 MGD and greater. They are not included for the 1-MGD flow because the capital cost of the turbines evaluated outweighs the power savings for flows less than 5 MGD.

Total Reverse Osmosis Seawater Desalination Costs

To compare the cumulative impact of some of the desalination process parameters and siting factors, a range of total costs for RO seawater desalination facilities are shown in Table 4-4. These costs are for an example facility treating seawater with an average salinity of 30,000 mg/L TDS that produces an average of 25 MGD of desalinated water. Most of the typical assumptions shown in Table 4-2 are used. Some of the parameters are modified to account for varying source water quality. The parameters from Table 4-2 that fluctuate

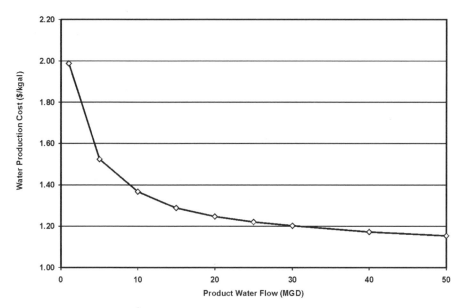

Figure 4-6 *Product water flow cost impact.*

are the recovery rate, which ranges from 40 to 60%; the flux rate, which ranges from 6 to 10 gfd; and the cleaning frequency, which ranges from once every 2 weeks to once every year. Other modifications are specific to individual portions of the desalination process and are explained below. The financial assumptions in Table 4-2 are used for all portions of the estimates.

Raw water supply includes the necessary intake structure, pumps, and piping to deliver seawater to the RO treatment plant. Raw water supply facilities on the low end include only minimal pumps and piping for a desalination plant that is co-sited with a power plant that has an adequate intake structure for use by the desalination plant. Raw water supply facilities on the high end include a large-intake structure with precautions to prevent impingement, an intake canal several thousand feet long, pumps, and piping.

The desalination process includes all necessary pretreatment, feedwater pumping, an RO membrane process system, and a cleaning system. The desalination process on the low end is for the treatment of an ideal source water that requires minimal pretreatment, allows the membranes to operate at around the maximum design flux rate and recovery rate, and does not require frequent cleaning of the membranes. The desalination process on the high end is for poor source water that requires extensive pretreatment, including coagulation and filtration, prevents the membranes from operating at a high design flux rate and recovery rate, and requires frequent cleaning of the membranes.

Concentrate disposal includes the necessary outfall, pumps, and piping to dispose of the RO concentrate to surface water. Concentrate disposal facilities

TABLE 4-4 Total Reverse Osmosis Seawater Desalination Cost Range

	Low Estimate			High Estimate		
	Capital Cost	O&M Cost	$/kgal	Capital Cost	O&M Cost	$/kgal
Raw water supply	$1,100,000	$200,000	0.03	$40,000,000	$2,000,000	0.54
Desalination process	51,000,000	6,200,000	1.09	105,000,000	15,000,000	2.48
Concentrate disposal	6,900,000	370,000	0.10	112,583,000	977,000	1.00
Delivery to demand center	17,382,000	300,000	0.17	205,336,000	2,840,000	1.95
Total	$76,382,000	$7,070,000	1.38[a]	$445,919,000	$17,817,000	5.97

Note. Cost is expressed in dollars per 1000 gallons of product water. Costs are for plants producing an average of 25 MGD of desalinated water. Costs are for reverse osmosis desalination of seawater with average salinity of 30,000 mg/L TDS. Each case is site-specific and costs can vary beyond these ranges. O&M, operations and maintenance.
[a] The total low estimate represents an idealized condition that could not actually occur on any single site.

on the low end include only minimal pumps and piping for a desalination plant that is co-sited with a power plant that has an adequate outfall for use by the desalination plant. Concentrate disposal facilities on the high end include pumps, piping, and diffuser for an open ocean discharge into waters a minimum of 30 ft deep.

Delivery to demand center includes the necessary pumps, piping, and water storage tanks for supply of the desalinated water to the distribution system. Delivery to demand center on the low end includes a 13-MGD storage tank with pumps and pipes for delivery 1 mile to the distribution system. Delivery to demand center on the high end includes a 13-MGD storage tank with pumps and pipes for delivery 140 miles to San Antonio.

EXAMPLE SEAWATER DESALINATION SITES ON THE TEXAS COAST

Sites were chosen to present example costs for a complete seawater desalination water supply on the Texas coast. Facilities were assumed to supply 25 MGD of desalted water. The example presented below is a facility co-sited with a power plant in Corpus Christi. Financial and other assumptions given in Table 4-2 were used except where stated in the example. Site-specific water quality and physical conditions for the location were used to the extent possible.

Example: Corpus Christi

The seawater desalination facility for Corpus Christi was assumed to be located next to the Barney M. Davis Power Station between Laguna Madre Bay and Oso Bay in south Corpus Christi. Figure 4-7 shows the location for this example. Davis is a once-through cooling water power plant with an existing reported cooling water flow of 467 MGD. Cooling water is diverted from Laguna Madre Bay and returned to Oso Bay. Engineering assumptions for the Davis seawater desalination example are shown in Table 4-5.

The estimate assumes that the power plant seawater intake is utilized to obtain the RO treatment plant feedwater. Drawing the source water from the power plant discharge eliminates the need to draw additional flow from the bay for cooling water and supplies feedwater with an increased temperature that is beneficial for the RO process.

Preliminary data indicate that there may be insufficient flushing in Oso Bay and the other surrounding bays for discharge of the RO concentrate. Therefore, for this estimate a separate RO concentrate disposal outfall is included to pipe the RO concentrate to the open Gulf. The outfall crosses Laguna Madre Bay and Padre Island and extends into the Gulf to be diffused in water over 30 ft deep. The concentrate disposal assumptions used in Figure 4-4 were applied, including the assumption that half of the concentrate pipeline will be located through sea grass beds and appropriate mitigation will be required.

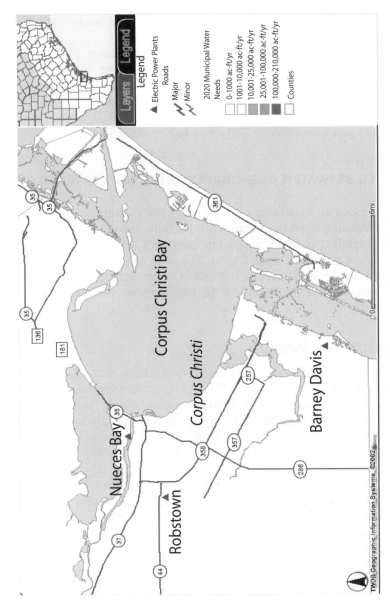

Legend

Electric Power Plants
▲

Roads
- Major
- Minor

2020 Municipal Water Needs
- 0-1000 ac-ft/yr
- 1001-10,000 ac-ft/yr
- 10,001-25,000 ac-ft/yr
- 25,001-100,000 ac-ft/yr
- 100,000-210,000 ac-ft/yr
- Counties

Nueces Bay

Corpus Christi Bay

Corpus Christi

Robstown

Barney Davis

TWDB Geographic Information Systems, ©2002

0   6mi

Figure 4-7   *Example: Corpus Christi.*

TABLE 4-5 Seawater Desalination at Barney M. Davis Power Station: Engineering Assumptions

Parameter	Assumption	Description
Raw water salinity	33,000 mg/L	Intake from power plant at Laguna Madre Bay
Raw water total suspended solids	40 mg/L	
Finished water chlorides	100 mg/L	Existing median at Stevens Plant is about 120 mg/L
Product water flow	25 MGD	
Concentrate pipeline length	10 miles	Diffused in open gulf in over 30 ft of water
Treated water pipeline length	20 miles	Distance to Stevens Plant or port industries
Feedwater pumping head	900 psi	
Pretreatment	High	Coagulation, media filtration, and chemical addition
Posttreatment	Stabilization and disinfection	Lime and chlorination
Recovery rate	50%	
Flux	8 gfd	Rate product water passes through membrane
Cleaning frequency	6 months	Membranes cleaned once every 6 months
Membrane life	5 years	Membrane elements replaced every 5 years

Water treatment parameters are estimated based on available water quality data for Laguna Madre Bay near the power plant intake. Coagulation and media filtration is included, along with the other standard pretreatment components (cartridge filtration, antiscalant and acid addition). Included sludge handling consists of mechanical sludge dewatering and disposal to a nonhazardous waste landfill. A product water recovery rate of 50% was used for this example. This is a lower recovery rate than the 60% reported for the Tampa Bay Water project. The lower recovery rate is anticipated due to the higher average salinity of the Laguna Madre Bay at 33,000 mg/L TDS as compared to the water source for the Tampa Bay Water project at 26,000 mg/L TDS.

Land acquisition includes 20 acres for the desalination plant and 97 acres for the desalted water storage tank and transmission pipeline. No land acquisition is included for the concentrate disposal pipeline but surveying costs are included. A 13-MG water storage tank and water transmission pumps and pipeline are included to transport the product water 20 miles to either the Stevens plant to blend into the city system or to distribution lines supplying industries along the ship channel. Posttreatment stabilization and disinfection are included.

Table 4-6 shows the cost estimate summary for seawater desalination at Barney M. Davis Power Station. The estimated total cost at 100% utilization

TABLE 4-6   Seawater Desalination at Barney M. Davis Power Station: Cost Estimate Summary

Item	Estimated Costs	
	100% Utilization	85% Utilization
Capital costs		
Source water supply	$800,000	$800,000
Water treatment plant	72,000,000	72,000,000
Concentrate disposal	32,000,000	32,000,000
Finished water transmission	20,000,000	20,000,000
Total capital cost	**$124,800,000**	**$124,800,000**
Other capital cost		
Engineering, legal costs, and contingencies (35%)	$43,680,000	$43,680,000
Land acquisition and surveying	2,100,000	2,100,000
Environmental and archaeology studies and mitigation	6,900,000	6,900,000
Interest during construction (6% for 2.5 years)	18,720,000	18,720,000
Total project cost	**$196,200,000**	**$196,200,000**
Annual costs		
Debt service (6% for 30 years)	$14,254,000	$14,254,000
Operation and maintenance		
Source water supply	200,000	200,000
Water treatment plant (except energy)	8,000,000	6,900,000
Water treatment plant energy cost	4,300,000	3,700,000
Concentrate disposal	700,000	650,000
Distribution	700,000	650,000
Total annual cost	**$28,154,000**	**$26,354,000**
Available project yield (AFY)	**28,004**	**23,803**
Annual cost of water ($ per acft)	**$1,005**	**$1,107**
Annual cost of water ($ per 1000 gallons)	**$3.08**	**$3.40**

of $3.08 per 1000 gallons of product water is about 45% higher than the lowest proposal received for the Tampa Bay Water desalination project. The estimated increased costs for this project are primarily the result of higher source water salinity and additional costs for the concentrate disposal pipeline and diffuser system. The total product water cost at 85% utilization is estimated at $3.40 per 1000 gallons.

Permitting of this facility will require extensive coordination with all applicable regulatory entities. Use of the existing power plant intake should facilitate permitting for the source water because no additional water is to be drawn from the bay. However, permitting the construction of the concentrate pipeline across Laguna Madre and Padre Island and construction of the ocean outfall will be major project issues.

CONCLUSIONS

Analysis of the Tampa Bay Water desalination project and siting conditions on the Texas coast indicate that a seawater desalination project on the Texas coast may be economically feasible. At just over $3.00 per 1000 gallons of product water, the cost developed for the example site at Corpus Christi is about 50% higher than the lowest proposal received for the Tampa Bay Water desalination project. However, the example cost is not far above the current range for other water supply projects to provide large quantities of potable water to these portions of the Texas coast. Also, desalination costs have decreased by a factor of 2 or 3 in the last ten years and may continue to decrease in the future, although further cost decreases will most likely proceed at a much slower pace.

Additional information will be needed once a site has been identified as a potential seawater desalination location. The Tampa Bay Water desalination project provides an example of the kind of information required to reduce uncertainty about the suitability of a particular location for a desalination facility. Tampa Bay Water obtained several environmental reports and studies that helped establish the feasibility of a desalination plant disposing of concentrate to a Florida bay or the Gulf of Mexico. Reports included an analysis from the U.S. Geologic Survey on the water transport in Lower Hillsborough Bay, Florida. This USGS report helped establish that there is most likely sufficient flushing in the bay to allow discharge of the desalination concentrate without salinity buildup. If concentrate discharge to a Texas bay is pursued, a similar analysis is needed to determine the water transport characteristics of the Texas bay that is being considered as receiving water for concentrate. Tampa Bay Water also commissioned a report titled "Impact Analysis of the Anclote Desalination Water Supply Project." This report focused on the potential environmental impacts associated with (1) the discharge of desalination plant concentrate to the coastal estuary of the Anclote Sound and (2) the intake of ambient surface waters for potable water production. These are the two primary environmental concerns that will need to be addressed for a Texas coastal desalination facility.

The above-mentioned Tampa Bay Water siting evaluations are only the ones performed prior to receiving best and final offers from the developers. Additional detailed studies will be required once a site has been settled upon to ensure that all regulatory requirements are met. The selected Developer for the Tampa Bay Water project was required to perform all additional studies required to obtain permits for the seawater desalination facility.

ACKNOWLEDGMENT

The authors thank the Texas Water Development Board, Nueces River Authority, Central Power & Light Company, City of Corpus Christi, and the San Patricio Municipal Water District for providing funding and participation for this research. A special thanks for the participation and guidance of the project steering committee including JD Beffort, Texas Water Development Board; Greg Carter, Central Power & Light Company; Ed Garana, City of Corpus Christi; Don Roach, San Patricio MWD; and Bill Hartley, BHP Engineering

& Construction, Inc. The contributions of Jim Vickers, Malcolm Pirnie, and Mark Farrell, Water Resources Associates, are greatly appreciated.

REFERENCES

1. Black, Bryan, and Mark Graves. Desalination for Texas water supply. Texas Water Development Board, August 2000.
2. *Membrane & Separation Technology News*, October 1999.
3. Levesque, Victor A., and K. M. Hammett. Water transport in Lower Hillsborough Bay, Florida, 1995–96. *U.S. Geological Survey Open-File Report 97-416*, Tallahassee, FL, 1997.
4. Bianchi, Pennock, and Twilley. *Biogeochemistry of Gulf of Mexico Estuaries*. New York: Wiley, 1999.
5. PBS&J, Inc. Impact analysis of the Anclote Desalination Water Supply Project. Prepared for Tampa Bay Water, November 1998.
6. Response to Best & Final Offer Seawater Desalination Water Supply Project for Tampa Bay Water. Stone & Webster, 1999.
7. Mickley, M., et al., *Membrane Concentrate Disposal.* AWWA Research Foundation and American Water Works Association, 1993.
8. Department of Energy, 1981.
9. Ramen, Raghu. PB-KBB Houston, TX. Personal communication, March 2000.

Ultrafiltration as Pretreatment to Seawater Reverse Osmosis

Merrilee A. Galloway
John G. Minnery,
Ionics, Incorporated, Watertown, Massachusetts

Reverse osmosis is the technology of choice for seawater desalination today. To ensure optimum performance of SWRO (seawater reverse osmosis) systems, appropriate pretreatment must be selected. While ultrafiltration has been used for many years as pretreatment to reverse osmosis (RO) for industrial applications, recent advances in ultrafiltration (UF) make this a potential alternative to conventional pretreatment for SWRO systems. This case study discusses ultrafiltration membrane systems and their advantages for SWRO pretreatment over more conventional technologies, and it uses pilot data to illustrate the benefits of ultrafiltration.

SWRO PRETREATMENT

The need for appropriate pretreatment to ensure optimum performance of SWRO systems is well documented. Suppliers of RO membranes usually require the SDI_{15} (15-min silt density index) of the feed to the RO to be less than 5, and generally recommend that the SDI_{15} should be less than 3. This minimizes the problems caused by suspended solids blocking the brine spacers in an RO membrane module. Many SWRO systems desalinate seawater from a beach well intake. These seawaters tend to be low in suspended solids, so, generally, it is possible to achieve an SDI_{15} of less than 3 with conventional pretreatment. For these applications, the use of ultrafiltration would not be justified. However, SWRO systems with open intakes can be much more challenging, especially in industrial areas such as ports, due to high levels of suspended solids.

Many SWRO systems operate successfully with well designed, maintained, and operated conventional pretreatment systems. For example, the author's company owns and operates a plant in California that has run for eight years without the RO membranes needing to be cleaned.[1] However, upsets in the performance of a conventional system can lead to solids causing excessive brine spacer plugging and increased pressure drop on the concentrate side of the membrane. If chemicals are overdosed or the wrong chemicals are used in a conventional pretreatment system, this can cause a sometimes irreversible increase in the transmembrane pressure (TMP). Increased power consumption, increased chemical cleanings, reduced membrane life, and overall increases in the RO system's operating and maintenance costs are the results of the occurrences described above.

CONVENTIONAL SEAWATER RO PRETREATMENT OPTIONS

Seawater must be treated before it reaches the RO unit to remove suspended solids.[2] Typically, multimedia filters are used to effectively remove solids. The filters are designed to operate at a loading rate consistent with the overall plant design.

The suspended solids are removed from the water into the filtering media by a combination of mechanisms, including straining, interception, impaction, sedimentation, flocculation, and adsorption. A majority of the suspended material in the raw seawater is removed during this first filtration step. As filtration proceeds, the filters accumulate captured particles. This causes clogging and is generally monitored by the increase in head loss across the filter. The filters are cleaned, one at a time, by periodic backwashing. In backwashing, clean water at a high flow rate is introduced at the bottom of the filter, expanding the media and washing collected solids out the top. Backwashing typically takes place automatically once or twice a day.

Multimedia filters are followed by cartridge filters. The cartridge filter system is used to remove fine suspended matter from water, typically down to 5 μm in size. A cartridge filter consists of a filter housing and filter elements mounted to tube supports. Water enters the housing and flows through the filter elements. The suspended solids are trapped in the fine fibers of the filter. The particulate matter suspended in the water is retained in the filter elements and the elements are replaced periodically. Cartridge filter elements are not reusable and all cartridges in an array have to be replaced at one time for maximum efficiency.

Other SWRO pretreatment schemes are used in some circumstances. With very clean beach well intakes, cartridge filters alone may be sufficient to protect the RO membranes. For some open intakes where high suspended solids are experienced (e.g., where a river empties into the seawater or in a port), additional pretreatment steps, such as clarification, may be required.

UF FOR SEAWATER RO PRETREATMENT

Given the issues associated with some of the more conventional methods of pretreatment mentioned above, membrane filtration is now being considered as pretreatment to SWRO, especially for some of the most tough-to-treat open intake seawater sources. UF has been found to have many benefits in as pretreatment to RO systems.[3] Many of these benefits may also apply to SWRO systems, as discussed below.

1. Membrane systems take up less than 50% of the area of a conventional pretreatment system. This results in lower construction costs. This means that a membrane system may be more favorable where space is limited or where the costs of civil works are high.
2. Since these systems operate in a dead-end configuration, the power usage for these systems is very low. RO power consumption may also be reduced since the RO is provided with cleaner water, and it will not foul or plug as much.
3. Conventional pretreatment systems often use large quantities of chemicals. This may include lime softening to reduce turbidity, ferric or alum, and polymers for coagulation. The RO system may also require more chemical cleaning with a conventional system.
4. While the operating costs of membranes systems may be lower than those for conventional pretreament systems, the capital cost of ultrafiltration systems is generally higher. This obviously depends on how extensive the conventional pretreatment needs to be in a particular situation. When compared with multimedia filters, ultrafiltration systems are obviously more capital intensive. However, when compared to clarifiers and filters, the cost difference between membrane and conventional systems is not as significant.
5. Given the automatic operation of membrane systems, the resulting filtrate water quality is very consistent regardless of feedwater quality variations. When large fluctuations in feedwater quality are noted, the more sophisticated membrane systems will compensate by frequency of backwashing or chemical cleanings. While a conventional plant will require an operator to check the water quality and adjust chemical dosing accordingly every few hours, UF plants require operator attention for only a few hours a day.

Filtrate Water Quality

Table 5-1 shows the expected filtrate water quality from microfiltration (MF) and UF systems and compares with multimedia filtration. Given that the UF systems typically produce water of a quality with and SDI_{15} less than 2 (and often below 1), the RO system should be easier to maintain. The benefits of the UF system include removal of suspended solids to prevent fouling and blockage of the RO brine spacer, elimination of large quantities of chemicals from the pretreatment to avoid the risk of overdosing, and removal of bacteria to reduce biofouling.

HOLLOW FIBER MEMBRANE FILTRATION

Over the last five years, hollow fiber membrane systems have gained wide acceptance for surface water treatment for potable water production. By 1999, over 200 MGD (million gallons per day) of installed capacity was in operation.[4] For potable water applications, hollow fiber membrane systems can guarantee removal of bacteria such as giardia cysts and cryptosporidium oocysts because the integrity of the membrane system can be verified. Extensive application of hollow fiber membrane filtration for potable water production has led to the costs of this technology coming down to the point where it is sufficiently cost competitive to be evaluated for SWRO pretreatment.[5,6]

Hollow fiber membranes for water treatment may either be MF or UF membranes. The fibers are typically 0.5–1 mm diameter. Several thousand hollow fibers are bundled into a membrane element. At either one or both ends of the membrane element, the fibers are potted in epoxy. Feedwater can be fed either to the inside of the fibers, with filtrate leaving from to the outside of the fibers (inside-out), or from the outside of the fibers, with filtrate leaving from the inside of the fibers (outside-in). Membranes are manufactured from several different materials, depending on the membrane supplier. Typical membrane materials include polysulfone, PVDF, polypropylene, polyacrylonitrile, polyethylene, and polyethersulfone. Many systems mount the hollow fiber modules vertically. A more compact design mounts the membrane modules in horizontal membrane housings similar to RO vessels.[7] Due to the presence of very fine colloidal silica in many seawater sources, UF is likely to be a better choice than MF for SWRO pretreatment.

TABLE 5-1 Filtrate Water Quality

Water Quality	Multimedia	MF	UF
Turbidity	0.5 NTU	<0.1 NTU	<0.1 NTU
SDI_{15}	3–5	<3	<2

Note. NTU, nephelometric turbidity unit.

NORIT UF TECHNOLOGY

The author's company has undertaken several pilot studies have used to evaluate the performance of hollow fiber UF on seawater. The membranes used were Norit X-Flow S225UFM. These membranes have a nominal pore size of 50,000 dalton molecular weight cutoff. The membranes fibers are permanently hydrophilic, which means they are less fouling over time. The membrane is contained in an 8-inch-diameter element, which can be utilized in the standard 8-inch pressure vessels developed for RO technology.

These elements are mounted in a horizontal configuration, and operated in the inside-out mode where the dirty water is introduced to the inside of the hollow fiber, with the filtrate passing to the outside of the fiber. The filtered water is then collected in a central filtrate tube, which reduces the pressure drop on the permeate side and also allows for greater backwash efficiency. Norit's UF Technology is operated in a dead-end mode. Particulates are removed from the membrane surface by means of a physical backwash that forces the particulates out of the membrane pores and away from the surface of the membranes. The backwash may occur every 20 minutes to every few hours depending on the system and the feedwater source. Since the system operates in a dead-end mode, operating pressures are generally low (usually around 10 psi), and there is no recirculation stream requiring extra pumping power. Over time, the physical backwash will not remove some membrane fouling. Most membrane systems allow the feed pressure to gradually increase over time to around 30 psi and then perform a clean in place (CIP). CIP frequency might vary from 10 days to several months. The alternative approach used for these pilot tests, is to use a chemically enhanced backwash (CEB), where, on a frequent basis, chemicals are injected with the backwash water to clean the membrane and maintain system performance at low pressure without going off-line for a CIP.[7]

Figure 5-1 shows an example of the change in TMP during the process. "A" depicts the normal filtration mode, "B" the backwash mode, and "C" a CEB. For any a given water quality and temperature, the duration A is merely a function of flux (flow rate per square meter of membrane). Hence, for a given interval between backwashes and CEB with a maximum design rise in TMP, the only parameter that can be adjusted is the flux rate. Therefore, if it is desired to increase the time between backwashes, the flux should be reduced. This will result in an increase in the capital costs since it will be necessary to increase the number of membranes so the appropriate amount of water is produced. Chemical pretreatment (e.g., addition of a few ppm of ferric chloride) can be effective in increasing the allowable flux rate.

The drop in TMP after a backwash is mainly dependent on the stickiness of the solids that have been accumulated on the membrane surface and the force applied during a backwash. A backwash will always be performed so that all of the backwashable solids are removed from the plant. This is generally the best way to postpone a CEB. Yet again, there is a trade-off between

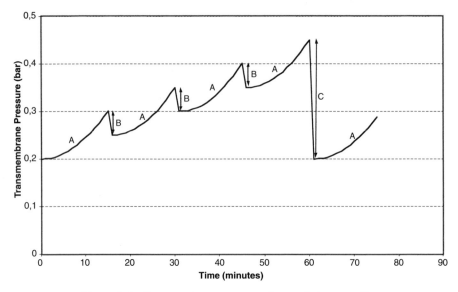

Figure 5-1   *Transmembrane pressure change during operation.*

the mechanical force applied during backwashing (i.e., the flow during back-washing) and the changing the stickiness of the solids layer.

A CEB (C) is performed whenever the TMP reaches a level determined during pilot testing, or at a set time interval. Chemicals used for seawater filtration are typically generic chemicals like sodium hypochlorite and nitric or sulfuric acid.

PILOT TESTING OF UF ON SEAWATER

Pilot testing has been undertaken in several sites around the world. The data presented in this paper have been collected from a pilot test that had been on-going in Trinidad for several months. The location for these tests is a large ship turning basin in Trinidad's Point Lisas Industrial Estate, on the Gulf of Paria. Due to the Gulf's location between Trinidad and Venezuela, adjacent to the mouth of Orinoco River, the water quality in the Gulf of Paria under-goes large seasonal variations, particularly in terms of salinity; conductivity oscillates on a seasonal schedule between approximately 22 and 52 mS/cm as shown in Figure 5-2. Turbidity is affected by the weather, seasonal changes, and the coming and going of ships—as tugboats position the ships they dis-turb the seafloor and temporarily increase turbidity as shown in Figure 5-3.

The pilot system employs a single element with 35 m^2 of membrane surface area. It was initially operated at a flow rate of 8.8 gpm, with alternative sodium hypochlorite and nitric acid CEBs. The choice of nitric acid was based

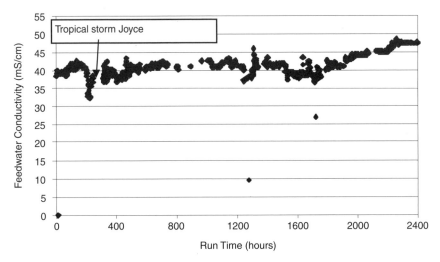

Figure 5-2 *Feedwater conductivity versus operating hours.*

on the recommendation of the membrane supplier based on results of pilot tests in the Arabian Gulf.[8] At 1220 hours of operation, the acid backwash was switched to sulfuric acid. Many seawater RO systems use sulfuric acid, so the use of sulfuric acid over nitric acid was evaluated to determine its acceptability. At 1862 hours of operation, 2 ppm of ferric chloride (as iron) was dosed to the UF feed to evaluate the effect on fouling. At this time the CEB sequence was changed to a regular sulfuric acid CEB, and a twice-

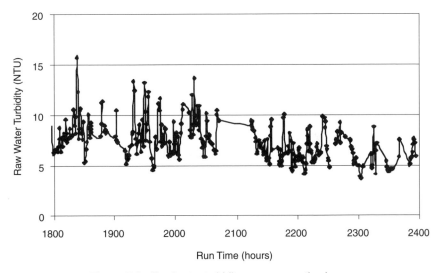

Figure 5-3 *Feedwater turbidity versus operating hours.*

weekly sodium hypochlorite CEB to control bio-growth. At 2174 hours of operation, the flow rate was increased to 13.2 gpm. Coagulation and micro-flocculation directly upstream of UF should allow it to operate at a higher flux without causing irreversible fouling.

RESULTS AND DISCUSSION

Figure 5-4 shows feedwater turbidity and UF product water SDI_{15} versus time. As expected with a membrane-barrier process, the product quality, as indicated by SDI_{15}, is not a function of the feedwater quality. The turbidity values are truncated above 20 NTU, since such values were found to be the result of debris accumulating in the light chamber of the online turbidimeter. This was verified as observed with a handheld turbidimeter: when a significant difference between the two instruments was suspected and confirmed, the online instrument was taken off line and cleaned. Over 95% of SDI_{15} measurements on UF product water were less than 3, and over 85% were less than 2. Most of the SDI_{15} values above 3.0 could not be repeated, and their cause is questionable; for example, they could result from contamination of the SDI apparatus or sample. Only at approximately 2130 hours of run time is there a cluster of four consecutively high values above 3.0 (3.2–3.4).

Figure 5-5 shows flow rate versus time and resistance (1/permeability) versus time. The relative stability of the resistance curve demonstrates that this mode of operation is sustainable and that it prevents irreversible fouling. A small increase in the resistance is observed after approximatley 2100 hours

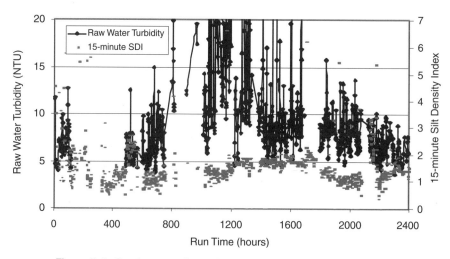

Figure 5-4 *Feedwater quality and product quality versus operating time.*

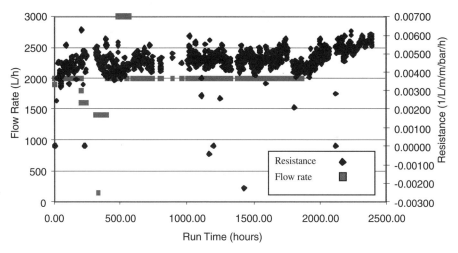

Figure 5-5   *Membrane resistance and flow rates versus operating time.*

of run time. This was expected with the addition of ferric chloride to the feed and with the 50% increase in flux.

CONCLUSIONS

UF product quality is consistent, regardless of upsets in the feedwater quality. This would ensure high-quality feed to a downstream SWRO system, with a pretreatment system that needs minimal operator attention to achieve this high quality. With ferric chloride in the feed, the system was able to operate at 50% higher flux without any signs of irreversible fouling. Sulfuric acid appears to be acceptable as an alternative to nitric acid for CEB on this seawater source. The next set of questions is whether the excellent SDI achieved by UF pretreatment will translate into superior performance of the SWRO system. Future work will evaluate the long-term benefits of utilizing UF as pretreatment to SWRO systems.

SUMMARY

The use of hollow fiber UF looks promising on difficult-to-treat open-intake seawaters. While hollow fiber UF may not be the answer for every type of seawater, it certainly is applicable in cases where there are large variations in the seawater quality that could cause difficult-to-handle upsets on conventional pretreatment. It is expected that the benefits of UF as pretreatment to

SWRO will soon be realized and that UF will be applied more widely in the months and years to come.

REFERENCES

1. Prato, T., E. Schoepke, L. Etchison, T. O'Brien, B. Hernon, K. Perry, and M. Peterson. Production of high purity water from seawater. Presented at ADA 2000 Biennial Conference & Exposition, South Lake Tahoe, Nevada, August 2000.
2. Meyer, S. Seawater reverse osmosis plant—energy recovery and capital cost reduction. Presented at The Changing World of Water & Power, Curacao, Netherlands Antilles, October 2000.
3. von Gottberg, A. J. M., and J. M. Persechino. Using membrane filtration as pretreatment for reverse osmosis to improve system performance. Presented at ADA 2000 Biennial Conference & Exposition, South Lake Tahoe, Nevada, August 2000.
4. American Water Works Assocation. Current issues in membrane applications and research. AWWA Membrane Technology Conference Preconference Workshop, February 1999.
5. Murrer, J., and R. Rosberg. Desalting of seawater using UF and RO—results of a pilot study. *Desalination* 118:1–4 (1998).
6. van Hoof, S. C. J. M., A. Hashim, and A. J. Kordes. The effect of ultrafiltration as pretreatment to reverse osmosis in wastewater reuse and seawater desalination applications. *Desalination* 124:231–242 (1999).
7. van Hoof, S. Semi dead-end ultrafiltration in potable water production. *Filtration + Separation* (January/February 2000).

Performance and Economic Evaluation of a 16-Inch-Diameter Reverse Osmosis Membrane for Surface Water Desalting

Christopher J. Gabelich
Tae I. Yun
Bradley M. Coffey
Metropolitan Water District of Southern California, La Verne, California

Robert A. Bergman
CH2M HILL, Gainesville, Florida

The Metropolitan Water District of Southern California (Metropolitan) supplies water to over 17 million consumers throughout southern California. Metropolitan's mission is to provide its service area with adequate and reliable supplies of high-quality water to meet present and future needs in an environmentally and economically responsible way. A major source of water for Metropolitan is from the Colorado River, which typically has 600–700 mg/L of total dissolved solids (TDS). The TDS of Colorado River water (CRW) may reach 750 mg/L in the future.[1] Recent studies have shown that

CRW causes approximately $95 million per year in damages to the public and private sectors for every 100 mg/L of TDS over 500 mg/L[1]—the U.S. Environmental Protection Agency's secondary, non-health standard. A planning goal at Metropolitan is to meet or exceed the 500 mg/L TDS secondary standard. One option to accomplish this goal is through desalination.

To make desalting economical at the large scale, Metropolitan is working in unison with other interested parties to develop and demonstrate new technology. The Desalination Research and Innovation Partnership was formed to evaluate and develop new desalting technology for the economical, large-scale treatment of surface water, municipal wastewater, brackish groundwater, and agricultural drainage water. The partnership consists of public water agencies, universities, private industry, and state and federal governmental agencies. It is anticipated that cooperation among these partners will improve desalting technology and broaden its use. One area under investigation is the potential capital cost savings associated with using large-diameter elements over commercially available 8-inch-diameter elements.

This case study presents test results using a reverse osmosis (RO) unit equipped with a single large-diameter (16 × 60-inch) element and a single conventional-diameter (8 × 40 inch) element. The large-diameter element has approximately five times more surface area than a standard 8-inch element. The large-scale RO unit was operated at 15 gal/ft^2/day (gfd) and 15% water recovery to simulate operational conditions potentially encountered in a full-scale, surface-water RO system with very good pretreatment. In addition, this study presents an analysis of potential savings associated with utilizing large-diameter RO elements based on new experimental data and published literature.

BACKGROUND

Metropolitan owns and operates five water treatment plants, three of which treat CRW. Each of the CRW plants (the F. E. Weymouth, Robert A. Skinner, and Robert B. Diemer filtration plants) are 520 million gallons per day (MGD) in size and use slightly different variations of conventional treatment (most typically rapid mix, flocculation, sedimentation, and dual- or tri-media filtration). The very large size of these plants is a result of favorable economies of scale associated with conventional water treatment systems. Conceptually, the 500-mg/L-TDS target could be met by treating a portion of the high-salinity CRW with RO and then blending the desalted water back with conventionally treated water (see Figure 6-1). However, given the high TDS rejection of current polyamide RO membranes (greater than 98.5% rejection of TDS) and split-flow treatment, the RO system at 85% water recovery would need to be at least 185 MGD (permeate flow) in size to lower the overall TDS from 750 to 500 mg/L. (The Colorado River Seven Party Agreement

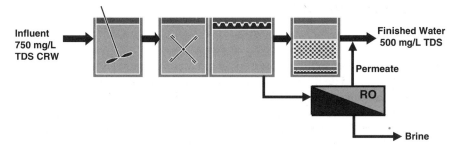

Figure 6-1 *Conceptual diagram of a 520-MGD, split-flow desalting facility.*

states that the salinity at the Parker Dam will not exceed 747 mg/L TDS [Colorado River Salinity Control Forum 1996].)

Historically, RO systems have not been used for very large plants because of the unfavorable scale-up costs associated with this technology. For example, a 150-MGD RO plant would require about 24,000 standard 8-inch-diameter by 40-inch-long, spiral-wound RO elements. These (smaller) RO elements are ideal for small to midsize plants, but are not practical for a very large system. Little additional economy of scale can be achieved for a very large facility needing so many individual pieces; the cost of constructing a 150-MGD plant would be roughly three times the cost of a 50-MGD plant. In contrast, larger RO elements should allow more convenient and economical construction of very large RO plants.

The use of large elements could have many other benefits, including a reduced number of seals (common sources of failures), a smaller footprint, and reduced maintenance requirements. A 150-MGD RO system would require 5000 large elements, compared to 24,000 conventional RO elements. Factors limiting the commercial introduction of this new technology, such as handling issues and membrane manufacturers' willingness to produce large-diameter elements, are also discussed.

When retrofitting an existing drinking water plant with RO, the full utilization of existing infrastructure and facilities greatly decreases the overall capital expenditures needed to meet water quality objectives. Previous research by Metropolitan has shown that when ozone is used as the preoxidant and the downstream dual- or tri-media filters are operated in a biologically active mode, a conventional treatment plant may serve as the pretreatment for an RO system.[2-4] Therefore, Metropolitan may be able to forgo building additional pretreatment processes (which would most likely be either microfiltration or ultrafiltration processes). Note that the treatment and disposal of the RO brine—a major stumbling block to the actual implementation of large-scale desalting in southern California—is not addressed in this article. However, the brine volumes with large or conventional RO elements would not differ and would not affect the comparison of capital costs.

EXPERIMENTAL METHODS

Pretreatment

Pretreatment was provided by Metropolitan's demonstration-scale plant in La Verne, California. The demonstration plant was located at the F. E. Weymouth Filtration Plant and could mimic the operational conditions seen during regular, full-scale, conventional treatment. While the demonstration plant was rated for 5.5 MGD, the plant was operated at only 3.0 MGD. Water was preozonated (0.95 mg/L ozone) in an over/underbaffled contactor to meet Surface Water Treatment Rule disinfection requirements (see Figure 6-2). Coagulant (2–4 mg/L ferric chloride) and cationic polymer (1.0 mg/L, polydimethyldiallylammonium chloride, Agefloc WT-20, CPS Chemical, Old Bridge, NJ) were fed prior to the flocculation basin. The water then passed through a sedimentation basin and an anthracite/sand dual-media filter (5.2 gal/min/ft^2 loading rate) that was operated biologically active. Finally, the water was irradiated with ultraviolet (UV) light (approximately 40-mJ/cm UV dose, Calgon Sentinal UV reactor, Calgon, Pittsburgh, PA) as part of another study being conducted concurrently. The UV irradiation may have provided a measure of protection against bio-fouling by reducing the influent loading of heterotrophic bacteria; however, previous bench-scale research did not show any protection against bio-fouling from UV treatment.[5]

Reverse Osmosis Unit

A 200-gpm RO unit was equipped with two pressure vessels operated in parallel (see Figure 6-3). The 8-inch-diameter by 40 inch-long pressure vessel housed a spiral-wound, thin-film composite, polyamide membrane element (Koch Fluid Systems TFC-4821ULP-400, San Diego, CA) with approximately 380 ft^2 of effective surface area. The 16-inch-diameter by 60-inch-long pressure vessel housed a new, experimental spiral-wound RO element (Koch Fluid Systems SE-16060ULP) with approximately 1950 ft^2 of effective surface area. The system was equipped with a fully automated control system (PanelView 1000, Allen Bradley, Milwaukee, WI) to collect all pertinent in-

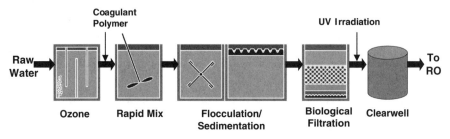

Figure 6-2 Schematic diagram of pretreatment process.

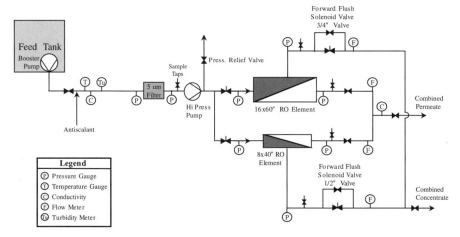

Figure 6-3   Schematic diagram of reverse osmosis test unit.

formation, such as flow, pressure, conductivity, temperature, and turbidity. Both RO elements were operated at flux and recovery levels that would be seen in a full-scale system operating at 85% recovery (15 gfd). Antiscalant (Permacare, Permatreat 191, Fontana, CA) was added prior to the 5-μm cartridge filter at a dosage of 1.6 mg/L. No pH adjustment was necessary, because the system was operated at low water recovery.

Modeling Approach

A hypothetical 185-MGD RO treatment plant was designed to produce low-TDS water. For the purposes of this paper, the cost of pretreatment (conventional treatment with ozone biofiltration) was assumed to be sunk and is not presented as part of the overall analysis. The cost information developed for this paper is solely for the 185-MGD, RO-treated sidestream (see Figure 6-1) and does not include brine handling or disposal. The location of the desalting facility was assumed to be at an existing conventional water treatment plant; therefore, the purchase of additional land was not needed. The amortization rate and period (6% and 20 years, respectively) were chosen based on established finance and planning practices at Metropolitan.

The RO plant design was based on the concept of small RO "building blocks," each consisting of 143 pressure vessels per train, using both 8- and 16-inch-diameter RO elements. This number was chosen based on an RO skid of 5.0 MGD using 8 × 40-inch elements, which approaches the practical upper limit for such an RO skid. Using 16-inch-diameter elements at approximately the same design flux, the individual RO skids increased in size to 16.8 MGD/skid. These building blocks were then replicated to treat the entire desalted flow. Thus, the RO plant costs would scale linearly, and economy-of-scale savings would be realized only for site work and systemwide controls.

The RO plant and membrane performance parameters used for the cost estimate are listed in Table 6-1.[6] Salt rejection was assumed to be equivalent between the 8- and 16-inch-diameter elements. However, the specific flux (flux divided by the net driving pressure) for the 16-inch prototype element was shown to be approximately 24% lower than that for a commercially available 8-inch RO element (see proceeding discussion). Assuming that no changes in capital equipment are needed, this would result in a 20% increase in applied feed pressure. It was also assumed that with further development, a second-generation, 16-inch-diameter RO element could be manufactured such that the specific flux is comparable to an 8-inch RO element. Applied pressure was calculated based on a 750-mg/L TDS influent at 64°F (18°C)— the annual median water temperature. Energy costs were calculated based on the applied feed pressures for both the first- and second-generation 16-inch RO elements (168 and 140 psi, respectively).

No RO brine treatment or disposal costs were assumed for this study; however, brine costs would be the same for large- or conventional-diameter RO systems. For a 185-MGD permeate capacity RO plant operating at 85% water recovery, the resulting brine stream would be approximately 33 MGD. This level of water loss in the arid Southwest would be unacceptable. Therefore, further brine treatment to increase the overall system water recovery would need to be instituted. Additionally, no brine lines to transport even a

TABLE 6-1 Plant Data for 185-MGD Permeate Capacity Reverse Osmosis System

	Value	
Parameter	8-inch RO System	16-inch RO System
Membrane element size	8 × 40 inches	16 × 60-inches
Membrane type	Ultra-low-pressure polyamide RO	Ultra-low-pressure polyamide RO
Membrane "effective" area/element	380 ft²	1950 ft²
Number of pressure vessels/train	143	143
Number of elements/ pressure vessel	6	4
Membrane permeate capacity/train	5.0 MGD	16.8 MGD
Number of treatment trains	37	11
Train footprint	1600 ft²/train	2200 ft²/train
Membrane unit recovery	85%	85%
Plant operating factor	98%	98%
Membrane flux	15 gal/ft²/day	15 gal/ft²/day
Influent TDS	750 mg/L	750 mg/L
Applied feed pressure	140 psi	168 psi (140 psi)
Permeate pressure	10 psi	10 psi
Feed water temperature	64°F (18°C)	64°F (18°C)

Note. Data in parentheses are for second-generation prototype elements with improved specific flux.

fraction of this brine stream are in existence; therefore, new brine lines would need to be constructed. Each of these additional costs would substantially increase the total desalting facility cost.

RESULTS AND DISCUSSION

Operational Data

Table 6-2 summarizes the operational data for the 8- and 16-inch elements. The specific flux over the initial 100 hours of operation for the 8-inch RO element was an average of 20% higher than the 16-inch RO element (see Figure 6-4). Over time, the difference in specific flux between the 16- and 8-inch RO elements increased to greater than 24% because of a higher fouling rate for the 16-inch RO element. The cause for the increased fouling rate for the 16-inch RO element is unknown at this time. Based on daily conductivity readings, the salt rejection for both elements was comparable (see Figure 6-5). Table 6-3 shows the percent rejection data for selected analytes. Overall TDS rejection for the 8- and 16-inch elements was 98.7 and 98.4%, respectively. However, data analysis for aluminum and iron were problematic because of their relatively low concentrations in the influent.

Differences in the membrane flux between the 8- and 16-inch-diameter elements were attributed to the following: (1) membrane material, (2) effective surface area, and/or (3) membrane leaf length. Small variations in the membrane material resulting from manufacturing process are expected and can account for a 5% difference in membrane performance. The effective surface area of the element is a critical value that directly affects the calculation of the specific flux. Although the actual membrane surface area that is used to manufacture an element is well known, the effective surface area of the membrane after gluing is uncertain. Typically, a 5% reduction in the

TABLE 6-2 Operational Data of Reverse Osmosis Pilot Unit

	Reverse Osmosis Element Size	
Parameter	8-inch	16-inch
Feed pressure (psi)	76.2	93.2
Differential pressure (psi)	7.9	5.8
Operating flux (gfd)	15.1	14.8
Process recovery (%)	14.3	14.0
Specific flux (gfd/psi)[a]	0.28	0.21
Normalized operating pressure (psi)[b]	65.6	81.7
Salinity rejection (%)	98.7	98.4
Energy usage (kWh/1000 gal)[c]	0.60	0.74

[a] Normalized to 25°C.
[b] Assume flux = 15 gfd, temperature = 25°C.
[c] Pump efficiency = 80%.

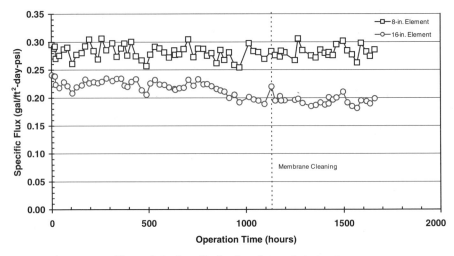

Figure 6-4 *Specific flux for pilot-scale test unit.*

effective surface area can be assumed after gluing of the membrane. However, because the 16-inch-diameter is a prototype element, assuming a 5% reduction in actual effective service area may be inaccurate. For calculations in this report, the effective surface areas of both the 8- and 16-inch elements were assumed to be 5% less than the manufacturers' specifications (380 ft^2 for the 8-inch and 1948 ft^2 for the 16-inch element).

The most significant difference between the two elements was determined to be the leaf length. The 8-inch element had a leaf length of 90 inches, and

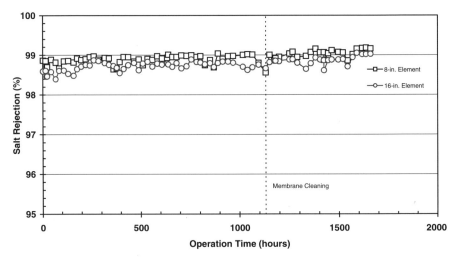

Figure 6-5 *Salt rejection for pilot-scale test unit.*

TABLE 6-3   **Water Quality Data**

Parameter	Reverse Osmosis Element Size	
	8-inch Rejection (%)	16-inch Rejection (%)
Inorganics		
Alkalinity	96.9 (5, 0.5)	97.2 (5, 0.2)
Total hardness	99.6 (5, 0.2)	99.5 (5, 0.0)
Total dissolved solids	98.7 (5, 0.8)	98.4 (5, 0.6)
UV absorbance at 254 nm	94.3 (5, 3.8)	93.9 (5, 3.4)
Calcium	99.7 (3, 0.2)	94.5 (3, 8.7)
Magnesium	99.8 (3, 0.0)	99.3 (3, 0.7)
Potassium	95.5 (3, 2.0)	96.7 (3, 2.5)
Sodium	97.6 (3, 0.2)	97.4 (3, 0.1)
Nitrate	93.5 (4, 1.2)	93.5 (4, 0.4)
Silica	97.6 (5, 1.0)	94.4 (5, 7.3)
Chloride	97.9 (5, 0.6)	98.2 (5, 0.7)
Sulfate	98.2 (5, 0.4)	98.3 (5, 0.5)
Fluoride	95.5 (5, 0.0)	95.4 (5, 0.0)
Trace Metals		
Barium	97.1 (5, 1.8)	98.3 (5, 1.9)
Aluminum	66.5 (5, 36.9)	67.6 (3, 22.4)
Iron	28.7 (5, 33.4)	35.4 (4, 40.8)
Strontium	99.2 (5, 0.9)	99.3 (5, 0.8)

Note. Data in parentheses indicate number of samples and standard deviation, respectively.

the 16-inch element had a leaf length of 130 inches. As the leaf length increases, the water must travel a longer distance to reach the permeate tube. As a result, the longer leaf length element requires more pressure than the shorter leaf length element at the same flux and recovery to drive the water through the semipermeable membrane to the permeate tube. It was hypothesized that the difference in the leaf length was the major contributing factor in the lower specific flux of the 16-inch element.

Membrane Cleaning

Both the 8- and 16-inch elements were chemically cleaned at 1130 hours of operation after a 17% reduction in flux had been observed for the 16-inch element (see Figure 6-5). The 8-inch element did not require cleaning, but both elements were cleaned so that they could be restarted at the same conditions. The membranes were cleaned with sequential applications of acidic solution (citric acid adjusted to pH 2.0–2.5) and caustic solution (equal parts Na-EDTA, sodium tripolyphosphate, and trisodium phosphate adjusted for pH 10.0–10.5). The flow rate during the cleaning cycle was 60 gpm for the 16-inch element and 30 gpm for the 8-inch element. Because of operational limitations of the cleaning-skid pump, the flow rate for the 16-inch element was not proportional to that of the 8-inch element (the manufacturer recommends a cleaning flow rate of 160 gpm for the 16-inch element).

After the chemical cleaning, a recovery in performance for the 16-inch element was not observed. The ineffectiveness of the chemical cleaning was attributed to the low crossflow rate through the 16-inch element during the cleaning cycle. In the future, the membranes will be cleaned separately to ensure that enough flow through the 16-inch element is available.

Cost Model

Table 6-4 shows the capital cost assumptions developed for this paper. Membrane costs ($650/element and $3380/element for the 8 × 40-inch and 16 × 60-inch RO elements, respectively) and pressure vessel costs ($1540 and $4950 for the 8- and 16-inch-diameter pressure vessels, respectively) were based on original equipment manufacturers' cost estimates.[7,8] These capital costs for RO include a $50 and $400 installation fee for 8- and 16-inch elements, respectively, and a $140 and $450 installation fee for 8- and 16-inch pressure vessels, respectively. All other costs were developed by CH2M HILL through either verbal quotes (e.g., membrane feed pumps), internal cost data (e.g., skid piping), or a proportion of the overall plant costs (e.g., elec-

TABLE 6-4   Capital Costs Assumptions

Item	8-inch RO System	16-inch RO System
Membranes		
Membrane	$600/element	$2,950/element
Installation	$50/element	$400/element
Pressure vessels		
Vessel	$1400/vessel	$4500/vessel
Installation	$140/vessel	$450/vessel
Skid piping	$300,000/train	$200,000/train
Support frame	$94,500/train	$62,500/train
Train instruments	$25,000/train	$25,000/train
Membrane feed pumps	$150,000/each	$300,000/each
Buildings		
Membrane train area	$100/ft²	$100/ft²
Other areas	$120/ft²	$120/ft²
Site development	$25,000/acre	$25,000/acre
Electrical	10% of equipment cost	10% of equipment cost
Plant controls	10% of equipment cost	10% of equipment cost
Construction contingency	25% of capital costs	25% of capital costs
Overall project contingency	20% of capital costs (including construction contingency)	20% of capital costs (including construction contingency)
Interest rate	6%	6%
Amortization period	20 years	20 years

trical components and plant controls). Operation and maintenance (O&M) costs were based largely on standard cost estimates (e.g., labor) or water quality requirements (e.g., chemical costs were estimated based on CRW water quality data to lower the influent pH to 7.0 and readjust the permeate pH back to 8.3) (see Table 6-5).

Table 6-6 shows a breakdown of the capital costs for a 185-MGD permeate capacity RO plant using both 8- and 16-inch elements. The overall capital cost savings for an RO system using 16-inch-diameter elements was 24%. This reduction in capital expenditures was largely a result not only of reducing the overall number of RO skids (thirty-seven, 5.0-MGD skids using 8-inch elements versus eleven, 16.8-MGD skids using 16-inch elements), but also of reducing the train piping, a cost savings of 80% per RO skid. The increased, rated skid capacity also resulted in substantial costs savings in RO skid instrumentation and membrane feed pumps for the overall plant. Not only were there fewer membrane feed pumps using 16.8-MGD skids, but each 19.8-mgd feed pump (16.8-MGD permeate flow divided by 85% water recovery) was proportionally less expensive on a cost per volume treated per day basis than a 5.9-MGD feed pump (5.0 MGD/85%).

Other capital cost savings associated with using large-diameter elements in greater capacity trains were a reduced plant footprint that resulted in reduced building costs (24% savings) and savings on systemwide plant controls and electrical equipment (28% reduction for each) (see Table 6-6). The only negative economy of scale was for the capital costs of installing the 16-inch RO elements. While the 16-inch membranes are competitive on the basis of cost per square foot of membrane ($1.53/ft^2 for 16 × 60-inch elements versus $1.58/ft^2 for 8 × 40-inch elements), the increased cost of installing the ele-

TABLE 6-5 Operation and Maintenance Cost Assumptions

Item	8-inch RO System	16-inch RO system
Labor		
Number of operators	30	30
Operator salary	$40,000/year	$40,000/year
Overhead	40% of labor	40% of labor
Chemicals		
Acid (93% H$_2$SO$_4$)	$0.04/lb	$0.04/lb
Scale inhibitor	$1.04/lb	$1.04/lb
Caustic (50% NaOH)	$0.14/lb	$0.14/lb
Cleaning chemicals	$0.015/kgal of permeate	$0.015/kgal of permeate
Cartridge filters	$5.40/cartridge	$5.40/cartridge
Other materials	$0.03/kgal	$0.03/kgal
Power	$0.06/kWh	$0.06/kWh
Pump efficiency	76%	76%
Motor efficiency	92%	92%
Plant operating factor	98%	98%
Membrane life	5 years	5 years
Membrane replacement	$600/element	$2980/element

TABLE 6-6   Breakdown of Capital Cost for 8-inch and 16-inch Reverse Osmosis System

	Value ($M)		
Parameter	8-inch RO System	16-inch RO System	Difference (%)
Membrane cost	20.6	21.3	3.4
Pressure vessels	8.15	7.79	−4.4
Skid piping	11.1	2.20	−80
Support frame	3.50	0.69	−80
Membrane feed pumps	8.39	5.00	−40
Other installed membrane train equipment	17.8	13.1	−26
Additional process items	11.3	11.3	0
Buildings	14.7	11.2	−24
Site development	0.63	0.63	0.0
Electrical	6.96	5.01	−28
Plant controls	6.96	5.01	−28
Other facilities	3.5	3.5	0
Construction contingency	28.4	21.7	−24
Overall project contingency	28.6	21.8	−24
Total capital	170.6	130.2	−24

ments ($400/16-inch element versus $50/8-inch element) negated this positive economy of scale. However, the cost for installing 16-inch RO elements is only an approximation and as more experience with large-diameter elements is gained, the cost to install large-diameter elements may decline.

Table 6-7 shows the total cost for a 185-MGD permeate capacity RO facility. When using the actual RO performance data collected during this study, the operation and maintenance (O&M) costs were 9% higher for the 16-inch RO system. The higher O&M costs resulted from the 20% higher applied feed pressure required by the 16-inch system (168 psi versus 140 psi for the

TABLE 6-7   Total Cost for a 185-MGD Reverse Osmosis System in Millions of Dollars

	Reverse Osmosis Element Size	
Cost Component	8-inch	16-inch
Energy ($M/year)	7.41	8.89 (7.41)
Labor ($M/year)	1.68	1.68
Chemicals ($M/year)	3.98	3.98
Membrane replacement ($M/year)	3.81	3.75
Miscellaneous ($M/year)	2.52	2.48
Annual O&M ($M/year)	**19.4**	**20.8 (19.3)**
Total Capital RO Cost ($M)	**170.6**	**130.2**
Annual RO Capital Cost ($M/year)	**14.9**	**11.4**
Total Annual RO System Cost ($M/year)	**34.3**	**32.2 (30.7)**

Note. Data in parentheses are for second-generation prototype elements with improved specific flux.

8-inch RO system) that resulted in a concomitant increase in energy consumption. However, when equal specific fluxes were assumed (0.28 gfd/psi), the O&M costs did not change significantly using either membrane size. This resulted in part in the same amount of water needing to be pumped at the same pressure regardless of membrane size.

A slight decrease in membrane replacement costs was observed, but this represented only a small fraction of the overall O&M costs. It may be argued that O&M labor may decline slightly using large-diameter elements because they have fewer O-rings, a common source of failures that require labor-intensive maintenance. Alternatively, using large-diameter elements that cannot be manually handled may increase the labor cost. Neither of these considerations were not used in determining the O&M labor component. An important issue that was not addressed in this paper is the loading and unloading of the membranes. A dry 16-inch-diameter element weighs approximately 200 lb and when wetted, an individual element can weigh over 300 lb. For these large-diameter elements to be used at the full scale, it will be important to design a method of easily removing and loading them that is not labor-intensive.

When equal RO element performance was assumed, the O&M fraction of the total plant cost was approximately 57 and 63% for the 8-inch element and 16-inch element membrane systems, respectively. Given that the O&M for a large-scale desalination plant did not change significantly when using second-generation, 16-inch-diameter elements, the overall RO plant costs, both capital and O&M, decreased by only 10% (see Table 6-7). However, when using first-generation elements with element inefficiencies, the overall RO plant costs decreased by only 6%.

CONCLUSIONS

The prototype 16-inch-diameter RO element evaluated during this study operated at 20% lower specific flux when compared to a standard, commercially available 8-inch-diameter RO element. Because of the experimental nature of the prototype 16-inch element, inefficiencies are expected. However, with future modifications in the leaf length, better estimation of the effective surface area, and improvements in the manufacturing of larger-diameter elements, it is anticipated that many of the suspected inefficiencies can be eliminated. Both elements were shown to have excellent rejection properties and removed greater than 98% of the influent TDS.

The use of large-diameter-capacity membrane elements is estimated to reduce large-capacity RO plant capital costs by nearly 24% and unit production costs by approximately 10%. The costs excluded waste concentrate disposal, but this cost would be the same using either RO element size. The reduction in capital expenditures was largely a result of reducing the overall number of RO skids, as well as reducing the train piping and support frames. The increased skid capacity also resulted in substantial costs savings in RO skid instrumentation and membrane feed pumps. Other capital cost savings associated with using large-diameter elements included reduced plant footprint, which resulted in reduced building costs (24% savings), and savings on systemwide plant controls and electrical equipment (28% reduction for each). Further research and development should be continued to improve the costs for large systems even more dramatically.

ACKNOWLEDGMENTS

Funding for this project was graciously provided by the California Energy Commission's Public Interest Energy Research Program. Special thanks to Warren Casey (Koch Membrane Systems, Medford, Oregon) and Doug Eisberg (Progressive Composite Structures, Vista, CA) for their help in putting together the membrane cost estimate. Additional thanks are extended to the entire Water Quality Laboratory staff at Metropolitan, who set up and maintained the research platforms and collected the data used during this study.

REFERENCES

1. Metropolitan Water District of Southern California and U.S. Department of the Interior, Bureau of Reclamation. *Salinity Management Study; Final Report.* Sacramento, CA: Bookman-Edmonston Engineering, 1998.
2. Bartels, C. R., C. J. Gabelich, T. I. Yun, M. R. Cox, and J. F. Green. Effect of pretreatment on reverse osmosis performance for Colorado River water desalination. *Proc.* 1999 IDA World Congress on Desalination and Water Reuse, San Diego, California.
3. Gabelich, C. J., C. R. Bartels, T. I. Yun, and J. F. Green. Comparing treatment strategies for surface water desalting. Presented at the 2000 AWWA Annual Conference, Denver, Colorado.
4. Gabelich, C. J., T. I. Yun, C. R. Bartels, and J. F. Green. *Nonthermal Technologies for Salinity Removal: Final Report.* Denver, CO: AWWA & AWWARF (2001).
5. Mofidi, A. A., C. R. Bartels, B. M. Coffey, H. F. Ridgway, T. Knoell, J. Safarik, K. Ishida, and R. Bold. Ultraviolet irradiation as a membrane biofouling control strategy. *Proc.* 1999 AWWA Water Quality Technology Conference, Tampa Bay, Florida.
6. CH2M HILL, Inc. Unpublished data prepared for Metropolitan (Dec. 2000).
7. Casey, W. (Koch Membrane Systems). Personal communication, December 2000.
8. Eisberg, D. (Progressive Composite Technologies). Personal communication, December 2000.

Case Study 7

Membrane Integration for Seawater Desalting

Charles (Chip) Harris

Advanced Membrane Systems, Inc., Yorktown, Virginia

As the need for potable water from nonpotable sources increases around the world, so does the need for a reduction in the cost of producing this water. Of particular concern are the "life-cycle" costs, such as operational labor costs, power costs, membrane replacement, and consumable costs. The traditional designs for seawater reverse osmosis (SWRO) membrane systems, while potentially capital cost effective, are also a stumbling block for the reduction of the eventual operational costs resulting from ineffective pretreatment and, quite often, poor operational perspectives.

The idea of integrating different membrane technologies into a single entity is not new. It has been tried, with moderate success, since the initial development of microfiltration membrane technology. The largest problems seemed to be with the selection of the membrane technology available and their incompatibility with the source water problems. The use of spiral wound ultrafiltration systems in conjunction with brackish reverse osmosis or nanofiltration is fairly common in the ultrapure industry. These systems, however, do not suffer from the greater feed source contamination found in seawater applications.

A new approach is now using a more effective hollow fiber ultrafiltration membrane technology that does not need some of the complex equipment associated with other designs, thereby simplifying the operational characteristics of the system. It also provides a more effective barrier for the microbial contamination that plagues many SWRO designs. This case study examines the design of such a system, considering the capital cost impact as well as the benefits in regards to long-term operation and maintenance costs. Other potential savings, such as power consumption reduction and equipment reduction, are also addressed.

TRADITION OR PERDITION?

The traditional methods of pretreatment for seawater membrane systems have been discussed time and again as to their frailties and drawbacks. The use of media filtration in this area, while often posing serious threats to the membrane portion of the system, continues to be used and designed into many systems. This type of filtration cannot provide adequate barriers to the often unstable nature of the source water. Also, the opportunity for enhanced biological growth inside these filters means a greater opportunity for membrane failure.

To control bio-fouling in these systems we also have the time-honored practice of chlorination and dechlorination. This method of decontamination can, if not properly applied, cause long- and short-term damage to even the best-designed membrane systems. The use of chlorine in the backwash water for the media filters is somewhat common for high biological loading systems. Once again, there exists a potential source for chlorine contamination of the membranes if there are any problems with the backwash or rinse cycles. All in all, it does seem that traveling the traditional road of pretreatment can lead your membrane system into "perdition."

THE ROAD LESS TRAVELED . . .

Experimentation in the use of ultrafiltration (UF) and microfiltration (MF) technologies for the pretreatment of seawater membrane systems began in the 1980s. Because of the availability of design being limited to mostly spiral configurations, the failure rate on many systems designed in that period was high.

One such small system, installed in 1982 for Sea World in Florida, suffered greatly due to the fouling tendencies of the source water. And this source water was immaculately clean in comparison to any surface source of seawater. The largest problem, it seemed, was that the membrane material had a "charge" to it that attracted very fine biological particulate matter and caused it to coagulate on the membrane surface. While these spiral designed filtration membranes were an apparent poor choice for the seawater applications, they had (and have) excellent potential for waters with much lower fouling potentials and quickly found lasting success in the industrial markets.

In the 1990s, great advances were made in both membrane chemistry and element design and exploration into the use of hollow tube or fiber designs for seawater pretreatment. This led to the popular term of "integrated" membrane systems, due to the integration of differing technologies into a single system. And here is where we now have many different choices.

FOR WANT OF A MAP . . .

Which direction shall the system take? There are a variety of roads to choose, from outside-in to inside-out flow direction. There is the dilemma of cross-flow versus dead-end flow operation. We have the question of backwashes, such as how often and how long. And, of course, the ultimate question of direction: how much will the trip cost? This author believes that the shortest distance between any two points is a straight line. That is to say, the best combination of the above choices for seawater applications could appear to be the use of inside-out dead-end flow configured hollow fiber membranes for the pretreatment stage.

This configuration, as manufactured by Koch (Romicon), Hydranautics (Hydra-Cap), or Ionics/Norit (X-Flow), seems to have the best potential for both process success and life-cycle costs. When operated in the dead-end mode, a larger feed pump to accommodate high cross-flow velocities is not necessary, thus saving capital and power costs. And while the Koch and Hydranautics designs are single, stand-alone modules, the Ionics/Norit design has the advantage of being a "standard" 8- by 40-inch (or 60-inch) membrane cartridge suitable for installation of multiple modules (up to four) into an industry standard pressure vessel. For the purposes of this study the Ionics/Norit example is used.

A BRIDGE ACROSS AN OCEAN . . .

While this UF membrane design has only recently been introduced here in the Americas, it has been in use in Europe since the mid 1990s for potable production from surface water sources. Because this membrane provides a "bacteriological barrier" with up to 6–8 log removal, the potential for the use of this membrane as a pretreatment for seawater membrane systems is great (see Table 7-1).[1]

TABLE 7-1 Microbial Results

Parameter	Seawater	Post Prefiltration	Post UF Treatment
Coliforms	>50/100	>50/100	<1/100
E. coli	60/100	12/100	<2/100
F streps	30/100	8/100	<2/100
1-day plate counts	44	23	<1

When we are discussing any reverse osmosis (RO) process, seawater or otherwise, we face the basic truth that a true RO membrane does not function well as a filter. Therefore, it is important to reduce particulate fouling as much as possible to protect the membrane. The use of the silt density index (SDI) to measure this fouling is common throughout the industry. As you can see by the chart (see Figure 7-1), the SDI reduction is significant using UF pretreatment.[2]

By reducing the fouling potential we can significantly increase the flux and driving pressure of the RO membrane to increase system recovery. Potentially, a system recovery increase from 40 to 55% is possible. This significantly lowers energy costs and capital costs for similar sized systems. Although the use of UF for pretreatment raises potential overall capital costs, the overall life-cycle costs for the system are reduced due to less frequent membrane cleaning, less consumables costs through the elimination of the cartridge filters, and potentially longer membrane life.[2]

THE DESIGN OF THE ROAD . . .

What then are the elements of our "road" to a better treatment system for seawater? There are several basic elements of the integrated membrane system, and we will attempt to simplify the system design to highlight the major components.

The start of the system is with the raw water collection. The major differences here are the collection structure, dependent on surface intake, subsurface intake, or sea well collection. The seawater is then pumped to the

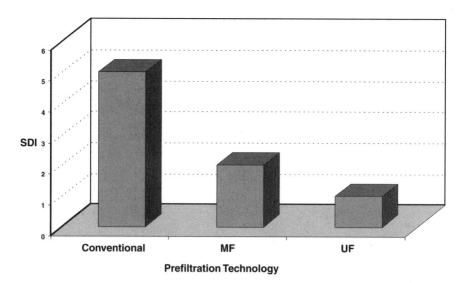

Figure 7-1 *Prefiltration SDI capabilities.*

membrane system, often via a collection tank of some sort. The membrane system starts with a roughing filter, usually a self-cleaning type rated between 200 and 500 μm dependent on suspected foulants. Filters of the spinning disk or high-velocity mesh types are the best for this service.

The next step is the UF portion, consisting of a number of hollow-fiber membrane modules, 8 by 60 inches and containing 35 m^2 of membrane area. These modules have a nominally rated flux of 70–100 L/h-m^2. Dependent on the size of the system, this portion could be broken up into multiple segments, each with its own flushing and isolation valves, to allow backwash of each segment individually while retaining service to the RO portion.

Next usually follows a break tank, to allow enough water for the backwash cycles while retaining RO service. These backwash cycles use UF filtrate to reverse flow the membrane every 20–30 minutes for 15–30 seconds. In addition, a chemically enhanced backwash (CEB) is conducted once every 4–24 hours, with a slightly stronger CEB cycle being introduced every 1–6 days.

All of the times mentioned above are variable and dependent on the source water. It is also important to remember that the CEB cycles usually contain a strong oxidant that needs to be completely rinsed from the UF elements prior to again feeding the RO portion. Also, a backwash neutralization tank might be needed to account for any CEB backwash disposal-related issues.

The final portion of the system is, of course, the SWRO membrane system. Even though we have stated above that higher recoveries are possible with the UF pretreatment, it is still important to consider the use of some form of energy recovery for this section. The higher recoveries mean higher pressure, and some mitigation of that energy requirement is preferred, if not mandatory.

COMING DOWN THE HOME STRETCH . . .

What other benefits can this combination of technologies have? A partial list of these benefits could be

- Modular construction—easily expandable
- Small footprint design
- Standardized technology using standard pressure vessel configurations
- Relatively long life of UF membranes
- Longer life of RO membranes
- Minimal installation costs
- Lower operational costs—less consumables
- Lower energy costs—higher RO recoveries

While all of the above are important, the first two benefits have the potential to make this idea of membrane integration very exciting. This integrated membrane technology could be combined with the idea of smaller, modular SWRO systems, capable of being installed almost overnight without the need for complicated civil structures, large building expenses or other major expenditures.

How might this be possible? By utilizing the idea of containerized membrane systems. In the past only the RO portion of a system was easily containerized, due to the weight and footprint of conventional pretreatment. But with the integrated membrane system, a new concept may emerge (see Figure 7-2).

In the conceptual drawing shown in Figure 7-2, all the elements necessary for a single, integrated system capable of a production rate of 250–350 m^3/d (60,000–90,000 gpd) of potable water from seawater are contained in four standard 20-ft sea containers. This system houses power generation, controls, membrane system, pumping systems, backwash storage, and permeate storage all in a four-element package that could be prefabricated and set up within a week. Variations of this theme could produce up to 1200 m^3/d (315,000 gpd in 40-ft containers) in much the same configuration for a variety of applications ranging from construction to resort use. The best features of this system are

- Self-containment
- Minimal footprint
- Ease of installation and commissioning
- Reasonable overall costs

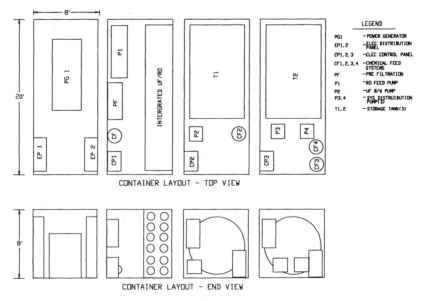

Figure 7-2 Potential container layouts.

An explanation of the last feature, reasonable overall costs, may be necessary. While the capital costs for this system estimate at $0.55 million US for the system shown, this is mitigated by the low overall energy consumption of under 3.6 kWh/m^3 attainable by the use of state of the art energy recovery technologies.[3] Another factor in this cost is the potential for the eventual resale of this type of system to another client. This further lowers the overall costs category.

THE END OF THE JOURNEY?

As we have seen in various demonstrations over the years, the best ways to make seawater desalination more feasible is to lower the costs and increase the reliability of these systems. And as we continue to draw water from coastal areas and arid sections, the need for potable water from seawater will only increase, and the less the consumer has to spend for this benefit the more rapidly it will come.

The newest innovations in this field are always exciting. And membrane integration, while not technically new, is a process that will continue to grow along with the water industry's need for innovation.

REFERENCES

1. Rosberg, R. Ultrafiltration, a viable cost-saving pretreatment for reverse osmosis. . . . *Desalination* 110:107–114 (1997).
2. Murrer, J., and R. Rosberg. Desalting of seawater using UF and RO. *Desalination* 118:1–4 (1998).
3. Hauge, Leif J. The pressure exchanger. *Desalination and Water Reuse* 9/1:54–60 (1999).

Case Study 8

Implementation of an Enhanced Multieffect Distillation Process for Seawater Desalination

J. Bednarski and F. Pepp

International Desalination Technologies, Pasadena, California

Presented at the AWWA Annual Conference, 21–25 June 1998, Dallas, Texas.

Starting in the late 1980s, intensive research and experimental activities were sponsored by the Metropolitan Water District of Southern California (Metropolitan) to develop a new seawater desalination process. These studies resulted in the conceptual design for a large-capacity state-of-the-art, multieffect distillation (MED) seawater desalination plant. This thermal-based process utilizes a 30-effect vertical tube evaporator (VTE) fabricated from double-fluted aluminum tubes, and is housed in a concrete containment vessel. This technology is referred to as the VTE-MED process.

In April 1998, a joint venture company, International Desalination Technologies (IDT), consisting of Parsons Overseas Company and IDE Technologies Ltd., and supported by Metropolitan and Reynolds Metals Company, completed detailed design of a 12.6-MGD (47,700 m³/day) demonstration facility that utilizes this innovative VTE-MED technology. Funding for IDTs work was jointly provided by the U.S.-Israel Science and Technology Foundation and Metropolitan.

This new VTE-MED technology offers several innovations over competing seawater desalination processes. The ones that led to the cost-effective production of desalinated seawater in large-capacity plants include (1) a vertical tube configuration of the process, (2) the use of aluminum double-fluted evaporator tubing, (3) an evaporator containment vessel (shell) fabricated from slip-formed concrete, (4) a high production yield of approximately 67%, (5) an economy ratio of approximately 23 lb of product water per pound of utilized steam, (6) relatively low energy consumption, as well as low consumables demand, and (7) a significantly smaller plant footprint.

The combined result of these innovations and the completed design effort for the 12.6-MGD plant is a VTE-MED seawater desalination technology that promises to produce water from large-capacity plants at less than $3.00 per 1000 gallons or approximately $980 per acre-foot ($0.80 per m^3).

PROJECT BACKGROUND

The Metropolitan Water District of Southern California (Metropolitan), established in 1928 to provide supplemental water to its member agencies, is one of the world's largest regional water suppliers. Metropolitan provides both treated and untreated water to approximately 16 million people in its 5200 sq mi (13,500 sq km) service area.

Metropolitan's interest in developing the potential of seawater desalination in Southern California goes back nearly 40 years. Most recently, Metropolitan has established itself as a well-recognized sponsor in the development of large-scale seawater desalination technology. Between 1988 and 1990 Metropolitan was involved in a number of feasibility studies to determine the most cost-effective method with which to desalinate seawater in large-capacity plants. Several of these studies included coupling a desalination plant with a fossil fuel or nuclear power station.

In mid 1990, Metropolitan's Board of Directors authorized a study to explore implementation of a 4.2-MGD (15,900 m^3/day) state-of-the-art seawater desalination demonstration plant. The study included preparation of a project plan, process design, and determination of water costs under various operating scenarios. To accomplish this work, Metropolitan assembled a group of selected design consultants from academia and the public and private sectors who had specialized expertise in various aspects of thermal desalination.

The process developed under this effort is referred to as the vertical tube evaporation (VTE) technology, a name that reflects the vertical tubular configuration of the distillation effects (evaporators). The process, while still at a concept-level of development, offered an economically feasible and technically attractive method of producing potable water from seawater in large-scale plants with capacities of up to 80 MGD (302,100 m^3/day). The comprehensive results of the aforementioned conceptual design effort are documented in Metropolitan Report No. 1084, "Seawater Desalination Plant for Southern California."[1]

With the completion of the Concept Design, Metropolitan pressed ahead with development of the VTE technology. In June 1992, work started on the preliminary design for the 4.2-MGD seawater desalination demonstration plant. An important phase of this work included the construction of a 2000-gpd (7.6-m^3/day) Test Unit evaporator in Huntington Beach, California. Following $2\frac{1}{2}$ years of in-house planning, design, and construction, Metropolitan placed the VTE Test Unit in operation in June 1995 for a two-year period. A variety of process and material performance data were subsequently collected and analyzed.

In 1995, an agreement was signed between U.S. President Bill Clinton and the late Israeli Prime Minister Itzak Rabin to jointly promote a number of "high-technology" projects of a research and development nature, whose successful development would be mutually beneficial to both countries. To achieve this goal, the U.S.-Israel Science and Technology Foundation (Foundation) was established to finance the materialization of such projects.

To pursue funding for a Seawater Desalination Demonstration Project, Parsons Overseas Company (Parsons) and IDE Technologies, Ltd. (IDE) formed a joint venture in May 1995. The joint venture, since named International Desalination Technologies (IDT), specifically submitted a proposal to the Foundation for funds to commercially develop the VTE-MED process. The IDT joint venture, together with Metropolitan and Reynolds Metals Company (Reynolds), would all be members of the team that would design, market, and construct the first commercial/demonstration VTE-MED plant with a capacity of 4.2 MGD.

In June of 1996 the Foundation formally selected the project for funding. Combined with matching funds from Metropolitan, the IDT design team officially began work on the project 1, July 1996.

The purpose of the IDT design/commercialization project was to

- Complete all salient experimental activities, especially verification of the evaporation heat transfer coefficient
- Engineer and design a demonstration plant of 4.2-MGD capacity in Southern California, or elsewhere
- Promote and market the new VTE-MED technology in countries where large-scale desalination projects were planned.

IDT assembled a consortium of companies chosen to complement each other in the development program.[2] The project participants and their specific areas of project responsibility are as follows:

- *Parsons Overseas Company* Overall organizational and management activities for the project, which encompasses the coordination of the various interface activities between the different intercompany task forces, as well as direct responsibility for the design of the concrete tower structure and the internal aluminum support framework for the evaporator.

- *IDE Technologies Ltd.*   Directly responsible for providing the process design and evaporator and seawater feed preheater tube bundle thermal design, including conceptual and detail design of the steam, brine, and product water distribution and collection systems.
- *The Metropolitan Water District of Southern California*   Project co-sponsor, as well as project subcontractor responsible for provision of testing services and support activities at the 2000-gpd Test Unit, and provided the design services for the "Balance of Plant" facilities at the demonstration plant.
- *Reynolds Metals Company*   Project subcontractor responsible for the detailed design of the evaporator and feed preheater tube bundles, tubesheets, and all other nonstructural interconnecting aluminum appurtenances required to provide a fully uniform and functional unit. Additionally, Reynolds fabricated one vertical effect tube bundle.

COMMENCEMENT OF THE DESIGN EFFORT: EARLY DESIGN BASIS MODIFICATIONS

During the early phases of Metropolitans VTE-MED conceptual process development, a 75-MGD (283,900-m³/day) prototype plant was chosen as the basis of design. Since a 75-MGD plant is composed of 18 sets of 30-effect evaporation strings, one 30-effect string would represent the design's smallest plant capacity, which is 4.2 MGD. It was therefore determined that the technology should be "tested" in a 4.2-MGD "demonstration plant" comprising one 30-effect evaporative string of tube bundles.

From the start of the design effort, however, it was recognized that it would be difficult to secure financing for a 4.2-MGD desalination plant, since in this size range the vertical MWD-MED process does not exhibit an appreciable economic benefit over competing desalination processes. Another area that presented problems was the structural stability of the 4.2-MGD unit. Because its diameter (20 ft) was small in comparison to its height (560 ft), it could not effectively be designed as one tower without the addition of an expensive external support structure.

Subsequently, it was decided by IDT to identify the smallest capacity, structurally self-supporting, VTE-MED plant that could be readily marketed. This plant would achieve three functions. First, it would serve as the basis of design for the Metropolitan/US-Israel-funded design effort. Second, the plant would demonstrate and prove out the merits of the VTE-MED process. Finally, this plant would be capable of economically producing desalted seawater. The capacity finally chosen for the demonstration plant was 12.6 MGD. This plant size requires three strings of evaporation modules and has a unit diameter of 40 ft. Such a configuration permitted the structure to be self-supporting with reasonable design and construction methods. When the demonstration plant capacity was upgraded from 4.2 to 12.6 MGD, it was projected that the production capacity would be representative of the majority of today's market demand, and additionally, would exhibit a total produced water cost below $3.00/1000 gal.

The 12.60-MGD Demonstration Plant Process Design

Seawater flows via gravity from the power station's existing intake channel, through the pretreatment systems, and into the VTE-MED unit. The seawater is pretreated in the following sequence:

- Chlorination, to destroy microorganisms and biological growth
- Strainers, to remove particulate matter down to 2 mm in size
- Ion trap, to minimize heavy ion attack on the aluminum internals of the evaporator
- Decarbonation by pH adjustment followed by vacuum degasification, to reduce the seawater alkalinity prior to entering the VTE-MED tower
- Deaeration, to eliminate noncondensable gases in the system
- Caustic addition, to control the pH, especially during acid overshooting
- Surfactant addition (optional), for heat transfer enhancement and to prevent buildup of hard scale on the evaporator tube surfaces
- Hard scale inhibitor (optional), to prevent any type of scale buildup on the evaporator surfaces especially hard calcium sulfate scale

The seawater next flows into the falling film condenser located in the lower portion of the evaporator tower. The seawater is then heated by steam from the 30th effect and falls into the seawater sump at the bottom of the tower. From there, it is pumped via the seawater feed pumps through the feed preheater (which is physically located in the core of the evaporator bundle), to the top of the evaporator (1st effect).

As the seawater feed travels up the preheater, it is successively heated by the condensing steam in each effect, until it is close to the evaporator temperature in effect No. 1. The vapors that condense outside the tubes of the preheater combine with the product water from other effects.

The heating media for effect No. 1 will be low-pressure steam from a backpressure turbine discharging at approximately 24 psia. This turbine will be fed from a 300-psig header in the power station. The 24-psig steam will condense on the external surface of the first-effect tubes, partially evaporating seawater inside the tubes. The condensate will be returned by gravity to the boiler house.

The seawater, now more saline (brine), and the produced vapor flow to the next lower effect. The brine flows downward and is collected on the upper tubesheet, before entering the tubes of the next effect. The vapors generated flow through the droplet separators located on the periphery of each bundle of tubes and then to the chest of next effect where condensation takes place. Between each effect, flashing of the brine and product water takes place due to the reduction of pressure between each successive effect. This process continues until the 30th effect is reached.

At the end of the process, the brine is pumped from the last effect (No. 30) of the tower either to the surfactant recovery system or directly to the outfall. The product water from all the effects and the condenser will be collected in the condenser and pumped out of the plant. The brine and the product water are pumped out of the tower by the brine and product water pumps, respectively.

The Surfactant Recovery System

The use of surfactant in the demonstration plant is optional. The benefits of utilizing surfactant are heat transfer rate enhancement and partial inhibition of hard scale formation, such as gypsum. To meet environmental regulations, a surfactant recovery system must be employed.

Experimental activities have shown that addition of surfactant ETA-27 at a rate of 10 ppm to the seawater will result in a concentration of 30 ppm in the outflowing brine. It is estimated that the normal environmental regulations will limit the brine surfactant concentration to approximately 10 ppm. Above this concentration, foaming in the outfall and in the sea might occur.

For the 12.6-MGD demonstration plant, a recovery system based on the flotation process has been selected. Recovery in the range of 65–70% is achievable, which reduces the brine surfactant concentration in the outfall from 30 to 10 ppm. The recovered surfactant can then be recycled back to the process thereby reducing the operating cost of this item by two-thirds.

THE HUNTINGTON BEACH DESALINATION TEST UNITS

Background and Objectives

In 1990, the Metropolitan Water District of Southern California began the design and construction of a pilot-scale, 2000-gpd Test Unit to verify conceptual design features of the VTE-MED process. This unit, located and erected on the site of Southern California Edison's Huntington Beach Power Plant, was designed by Metropolitan's engineering staff, with Reynolds fabricating the vertical tube bundles. Ancillary components of the Test Unit were provided by Metropolitan. Construction was completed during the summer of 1995, and the testing program started up immediately. Testing was completed in May 1997, and the unit was dismantled in June 1997.

Design of the Test Unit's individual components were identical to those of the demonstration plant, and therefore provided an experimental platform for validating Unit design parameters, equipment sizing, and mechanical layout configurations. The Unit also functioned as a development tool, providing data that assisted in the finalization of details and specifications subsequently incorporated into the demonstration plant design.

The primary objective of the Test Unit was the accurate measurement of heat transfer coefficients (HTCs) in the aluminum evaporator tubes. These data were deemed critical for establishing a basis for the design of the larger scale plant. These goals were accomplished through the use of the Short-Term Test Unit (STTU). The STTU was configured such that temperature and pressure conditions of the unit's two evaporative effects could be varied to replicate any two successive evaporative effects in a full-scale plant. The Test Unit was also designed to verify the ability of the VTE-MED process to meet a variety of water quality goals and long-term material performance objectives. These goals and objectives were obtained through use of the Long-Term Test Unit (LTTU). The LTTU was outfitted with instrumentation and controls that permitted long-term (week-to-week or month-to-month) operation of the VTE-MED process at single temperature and pressure conditions.

Test Results

The performance test program conducted at Metropolitan's Test Unit validated the most important criteria in the Concept Design for the VTE-MED.[3,4] The actual values obtained for the heat transfer coefficients of the aluminum evaporator tubes exceeded the levels originally predicted. Water quality of the product distillate met regulatory levels for potable water supplies. A summary of the anticipated heat transfer values and water quality related results that were obtained from the Test Unit are shown in Tables 8-1 to 8-4. The key findings from the Test Unit can be summarized as follows.

- *Operation of the Vertical Tube Configuration*   The two-effect pilot-scale STTU was capable of operating across the entire temperature range of a large-scale plant (i.e., from 90 to 230°F). Plant operation at all temperature test conditions was stable and repeatable. This consistent level of operation was proven with more than 70 test runs.
- *Evaporator Driving Force*   The temperature differences between the condensing (tube side) and the evaporating (shell side) were attained for all expected temperatures of operation and verifies the temperature driving force that is expected in a large-scale plant.
- *Heat Transfer Coefficients*   The most significant design feature of the proposed process is the enhanced heat transfer coefficients, which are attainable with the double-fluted aluminum evaporator tubes. The average HTC projected from Metropolitan's concept design was 1605 Btu/h-ft²-°F for the double-fluted aluminum tubes. The average HTC determined from the data collected at Metropolitan's Test Unit was in excess of 1700

TABLE 8-1   Heat Transfer Requirements for 30-Effect VTE Process (from Metropolitan's Concept Design)

Effect No.	Brine Temperature (°F)	Heat Transfer (Btu/lb ft°F)	Effect No.	Brine Temperature (°F)	Heat Transfer (Btu/lb ft°F)
1	230	1861	16	164	1725
2	226	1857	17	160	1694
3	222	1854	18	155	1652
4	218	1849	19	151	1608
5	214	1845	20	147	1561
6	210	1839	21	142	1511
7	205	1833	22	137	1455
8	201	1826	23	132	1396
9	197	1818	24	127	1331
10	192	1809	25	122	1260
11	188	1799	26	116	1181
12	183	1787	27	111	1093
13	179	1774	28	104	992
14	174	1760	29	98	875
15	169	1744	30	90	734

Btu/h-ft^2-$°F$. This extremely high heat transfer rate was attained without cleaning of the evaporator tube surfaces. In addition, a fouling factor of at least 0.0002 h-ft^2-$°F$ was used in the calculation of the heat transfer coefficient. Thus, the required heat transfer surface area projected in Metropolitan's concept design is conservative. This indicates that the performance ratio of 23.63 lb of distillate per 1000 Btu heat input will be met for a large-scale plant, as predicted in the Concept Design.

- *Distillate Production and Quality*   Distillate production rates for all runs averaged 90% of the Concept Design requirement. However, this figure includes many test runs before the distillate collection system was optimized to collect fully accurate measurements. Therefore, it is concluded that the Test Unit was capable of meeting the Concept Design requirement for distillate production. Distillate water quality, in terms of conductivity measurement, was consistently below 10 μmho/cm, well within design requirements.

TABLE 8-2   Criteria for Pretreatment of Raw Seawater

Constituent	Concentration Following Pretreatment
Bicarbonates	<10 ppm
Dissolved oxygen	<100 ppb
pH	6–7
Copper	<100 ppm
Iron	<100 ppm
Mercury	<100 ppm

TABLE 8-3 Aluminum Concentration Objectives and
Results

	Aluminum (ppb)
Maximum contaminant level allowed	1000
Secondary contaminant level allowed	200
Raw seawater concentration	5–40
Product distillate concentration	5–12

- *Scale and Corrosion Potential* Diagnostic observations of the LTTU following long-term operation at high-temperature operating conditions have been conducted. During plant operations, aluminum concentrations in the LTTU and STTU were determined by daily and weekly sampling. Table 8-3 summarizes the measured aluminum concentrations compared with raw seawater measurements and the maximum and secondary levels allowed by the State of California drinking water regulations. These data showed no significant loss of aluminum material in the product distillate or the brine. Extensive destructive testing of the aluminum tube bundles was conducted by Metropolitan's Corrosion Laboratory and Reynolds' Materials Development Laboratory after the Test Unit was dismantled in June 1997. No sign of scale or corrosion of the aluminum tubes or tube-sheets was found in any of the laboratory testing. These positive results established the integrity of aluminum materials of construction in high-temperature VTE-MED process conditions.

- *Process Pretreatment* The performance of the process pretreatment system, consisting of pH adjustment/control, heavy ion control, and decarbonation, is shown in Table 8-4. The Test Unit pretreatment system successfully controlled the evaporator seawater chemistry and avoided conditions that produce scaling or corrosion.

- *Product Water Quality* A sampling program was conducted to examine the quality of the raw seawater entering the units and the product distillate produced by the process. Water samples taken from the STTU and LTTU were shipped to water quality laboratories for analysis. The tabulated results of the analyses were compared against the maximum con-

TABLE 8-4 Seawater Pretreatment Objectives and Results

Constituent	Raw Seawater	Design Objective	Pretreated Seawater
Bicarbonates	180,000	<10,000	800
Dissolved oxygen	4,000	<100	<100
Copper	60	<100	60
Iron	40	<100	40
Mercury	ND	ND	ND
pH	8.2	6–7	6.8

Note. Units: parts-per-billion, except pH.

taminant levels allowed by the State of California Title 22 Drinking Water Regulations. This comparison showed that primary and secondary water quality standards were met for volatile organics, semivolatile organics, metals, and coliform bacteria. Total dissolved solids (TDS) in the raw seawater to the Test Unit average 34,000 mg/L. In the product distillate, TDS concentrations ranged from 2.5 to 20 mg/L, which is well below the maximum allowable contaminant level of 500 mg/L set forth in the State of California.

ALUMINUM EVAPORATOR EFFECTS

Evaporator Tubing

A key innovation of the VTE-MED is the use of double-fluted aluminum evaporator tubes as the primary heat transfer surface in the evaporator tube bundles. This is not the first use of aluminum tubes in seawater desalination. IDE has utilized aluminum evaporator tubes in their "standard" horizontal MED plants for nearly 30 years.[5] However, these tubes have a smooth-oval configuration, and IDE's horizontal MED process operates at relatively low temperatures (170°F maximum) compared to the VTE-MED (230°F maximum).

At the outset of the MWD-MED Program, Reynolds was requested to design an enhanced 2-inch-diameter tube for review by the designers of the process. In June 1991, a double-fluted tube design based on the original trapezoidal flute geometry and a preliminary extrusion practice was developed. The tube material utilized for this project is Reynolds' proprietary MG375 alloy. The alloy was developed over the years by Reynolds and is based primarily on their research and development work for the Office of Saline Water. The conceptual design was later modified to a radiused flute that increased the surface area and reduced the weight/foot of the section.

Initial extrusion trials of this section were conducted at Reynolds' plant in El Campo, Texas using MG375 and 3004 billets. The 3004 billets were used as control. The MG375 extrusion trials were deemed superior to the 3004 billets. The MG375 material met all preliminary requirements, and samples of the tubing were subsequently submitted to the desalination technical team for their review.

After further studies, the technical team requested that Reynolds design several alternative flute configurations. Nine additional 2-inch-diameter tube designs were made with variations of 36, 40, 48, and 52 flutes. At the request of the technical team Reynolds extruded one each of the 36-flute and 48-flute designs. Evaluations to date, however, including heat transfer data, have centered on the original 52-flute design, and it remains the design of choice.

Tubesheet Design

The following factors were considered in the design of the tubesheets:

- **Corrosion resistance**
- **Weldability,** with and without filler wire
- **Thickness,** as it affects joint integrity and bundle strength
- **Machinability,** as it affects the drilling speed and accuracy of the holes
- **Mechanical properties,** bundle strength, and proper fit between tubesheet and expanded fluted tube
- **Plate tolerance** as it affects fit-up during fabrication.

Five alloys (3003, 3104, 5052, 5154, and 6061) were candidates for the tubesheet. Alloys 3003 and 3104 were selected for their excellent corrosion resistance. The others were included to provide a wider selection of properties and/or performance. When it was apparent that both 3003 and 3104 provided suitable performance for the joint, the other alloys were no longer considered. Alloy 3104 was selected over 3003 based on favorable experience with a similar alloy, 3004, in the Wrightsville Beach facility, and due to the benefits of the magnesium in the alloy. Alloy 3003 does not contain magnesium.

The Tube-to-Tubesheet Joint

Since the desalination evaporator presents a potentially corrosive environment for aluminum (due to a combination of concentrated brine and high unit operating temperatures), efforts were focused on developing tube-to-tubesheet joints that reduced or eliminated crevices between the fluted tube and tubesheet, and on recommending alloys that would provide superior corrosion resistance.

A number of factors were considered in the selection of candidate joints, including elimination of crevices for improved corrosion resistance, cost of the joint, skill required to assemble the joint, equipment required, tolerances of the tube, tube sheet and machined hole in the tubesheet, reproducibility, and the process controls required to assemble the joint in production.

In the end, a roll-expanded joint with a silicone sealant filler was selected, because the roll expansion provides greater mechanical strength, and the silicone seals the joint and minimizes crevices. However, welding of the upper tube-to-tubesheet joint remains as a viable option.

PROJECT COST PROJECTIONS

Once the design effort had been completed, comprehensive material takeoffs and refined unit cost estimates were undertaken for the 12.6-MGD demon-

TABLE 8-5   Cost Ranges for VTE-MED Plants

Plant Capacity	Installed Capital Costs ($ million US)	Water Production Costs
12.6 MGD (47,700 m³/day)	70–90	$3.00–$3.50/1000 gal $980–$1140/acre-ft
30 MGD (113,580 m³/day)	125–155	$2.80–$3.20/1000 gal $920–$1040/acre-ft
80 MGD (302,100 m³/day)	300–350	$2.50–$2.75/1000 gal $815–$895/acre-ft

stration plant itself. These data were subsequently utilized as a base from which to develop a number of additional cost-estimate scenarios (cost projections) for VTE-MED facilities of varying capacities, in multiple locations in the world. These cost projections were performed for a 30-MGD (113,580-m³/day), 57-MGD (215,030-m³/day), and 80-MGD (302,100-m³/day) design. Cost projections, in the form of cost ranges for installed capital and water production costs, are summarized in Table 8-5 for the 12.6-, 30-, and 80-MGD plants.

During the development of the cost projections, it was observed that the final project cost for a VTE-MED plant is sensitive to site-specific factors and project-specific assumptions. The assumptions used to prepare the cost ranges in Table 8-5 are delineated in Table 8-6.

CONCLUSIONS

The vertical MWD multieffect distillation process has the potential to revolutionize the desalination industry, especially in the area of physical size ver-

TABLE 8-6   Site-specific Factors Utilized in Determining VTE-MED Cost Ranges

Capital Cost Criteria	Factor
Foundation costs	Deep caisson foundation with shallow groundwater along coastal Southern California
Intake/outfall costs	Existing power plant facilities utilized for seawater intake and brine discharge
Seismic requirements	Southern California site, Seismic Zone IV
Taxes/duties	Southern California site
Construction labor rates	Southern California labor market

Operating Cost Criteria	Factor
Interest rate	7%
Amortization period	25 years
Steam costs	$2.60 per 1000 pounds
Electric power costs	$0.068/kWh
Labor costs	Southern California labor market

sus overall plant capacity. Based on the results achieved to date on both the Metropolitan Test Unit, and in IDE's research facilities, it is anticipated that the cost of water in the very near future, will be substantially reduced over present-day commercially available desalination costs for similarly sized plants.

REFERENCES

1. Metropolitan Water District of Southern California, *Seawater Desalination Plant for Southern California-Preliminary Design Report No. 1084*, October 1993.
2. Pepp, F., et al. The vertical MWD MED (multi-effect distillation) process. International Desalination Association Biennial Conference, Madrid, Spain, October 1997.
3. Metropolitan Water District of Southern California. *Seawater Desalination Demonstration Program Summary Report No. 1124*, September 1996.
4. Bednarski, J., M. Minamide, and O. J. Morin. Test program to evaluate an enhanced seawater desalination process. International Desalination Biennial Conference, Madrid, Spain, October 1997.
5. Ophir, A., and J. Weinberg. *MED-(multi-effect distillation) desalination plants—a solution for the water problem in the Middle East.* International Desalination Association Biennial Conference, Madrid, Spain, October 1997.

Glossary of Desalting Terminology

Alkaline scale   Precipitated salts that dissolve under acidic conditions; usually calcium carbonate and magnesium hydroxide.

Alkalinity   The quantitive capacity of aqueous media to react with hydrogen ions. Bicarbonate, carbonate, and hydroxides in natural or treated water usually impart alkalinity.

Ambient temperature   The temperature of the surroundings, usually taken as 70°F.

Anion   A negatively charged ion in solution.

Anode   The positively charged electrode in a dc circuit.

Antiscalant   A chemical that inhibits the precipitation of sparingly soluble inorganic salts.

Aquifer   A geological formation, group of formations, or part of a formation capable of yielding a significant amount of water to a well or spring.

BAT   Best available technology.

Boiling point   The specific temperature and pressure at which the vapor pressure exerted by a liquid equals ambient pressure.

Brackish water   Saline water having a total dissolved solids concentration ranging from 500 to 10,000 mg/L.

Brine   Concentrate (reject) stream from a cross-flow membrane device performing desalination.

Bump   A liquid surge used in electrodialysis to displace gases formed at the anode.

Cathode   Negatively charged electrode in a dc circuit.

Cation   A positively charged ion in solution.

Cationic membrane   A membrane in electrodialysis with an overall negative charge that passes positively charged ions in an electric field and rejects negatively charged ions.

Cellulose acetate   An organic polymer used to make semipermeable membranes.

Chemical rejuvenation   A term indicating any of a number of in-place chemical cleaning methods to remove fouling material and scale or to recondition membranes.

Colloid   A small discrete solid particle, typically 0.1 to 0.001 mm in diameter, that is suspended in a solution.

Compaction   Compression of reverse osmosis membranes due to long-term exposure to pressure and temperature that results in a decreased water flux rate.

Concentrate   One of the output streams that has increased concentration of ions and particles over the feed stream.

Concentration polarization   The increase of the ion concentration over the bulk feed solution that occurs in a thin boundary layer at the feed side of the membrane surface.

Condensate   The liquid produced by cooling a gas to below its dewpoint.

Conductivity   The property of a substance (in this case ions and water) to conduct electricity.

Conversion   The amount of product water divided by the amount of feedwater; generally reported as a percentage.

Cpu   Chloropalinate units (color indicator).

DBP   Disinfection by-products.

D-DBP   Disinfectant disinfection by-product rule.

Deaerate   To remove dissolved air from a liquid solution.

Decarbonation   A process to remove carbonate alkalinity from feed water in the form of CO_2 gas.

Demister   A device that removes entrained liquid droplets from a gas stream.

Diffusion   The movement of a molecule in solution due to a difference in a concentration of that molecule in one place versus another.

Distillate   A liquid produced by condensation of vapors of that liquid, forming the product of a distillation process.

Distillation   A purification process in which a liquid is evaporated and its vapor is condensed and collected. For water treatment distillation is used as a desalting technique in such processes as multistage flash distillation, multiple-effect distillation, and vapor compression.

DOC   Dissolved organic carbon.

Droplet separator A device that removes entrained liquid droplets from a gas stream. Also see demister.

Eductor A device that uses a high-velocity liquid or gas stream flowing through a nozzle to produce a lower pressure (vacuum).

Effect A stage in a water distillation or evaporation system where steam or water vapor is condensed and its released energy used to boil water.

Electrode Any terminal that conducts electricity to or away from various conducting substances in a circuit.

Electrodialysis (ED) A desalination process using an electric field to transfer dissolved ions from a less concentrated to a more concentrated solution through ion-selective membranes to produce water low in total dissolved solids.

Electrodialysis reversal (EDR) An electrodialysis process that periodically reverses the flow of ions in solution by reversing the artificially produced electric field.

ESWTR Enhanced surface water treatment rule (U.S.).

Evaporation See distillation.

Falling film evaporation A type of heat exchange system where a thin film of liquid is brought to boiling as it flows by gravity over a heat exchange surface.

Feedwater Input or raw stream into the desalting process.

Finished water Water that has passed through a water treatment plant, such that all the treatment processes are completed, or finished. This water is ready to be delivered to consumers. It is also called product water.

Flash evaporation The sudden boiling of a liquid due to a pressure reduction.

Flux Flow of water through a semipermeable membrane; can be expressed as gallons of water per day per square foot of membrane area.

Fouling The reduction of water mass transfer by materials in the water typically caused by silts, colloids, and biological matter depositing on a membrane or heat transfer surface.

FRP Fiberglass reinforced plastic.

GAC Granular activated carbon.

HAA Haloacetic acids.

Hardness The summation of polyvalent cation concentration is water (normally calcium and magnesium ions); generally reported as mg/L calcium carbonate equivalent.

Hard scale or nonalkaline scale Calcium sulfate or other sparingly soluble materials that cannot be dissolved by acid.

Heat exchanger A device that allows thermal energy to be transferred from a high-temperature medium to a low-temperature medium.

Heat of vaporization   The amount of energy per unit mass required to convert a liquid to a vapor.

Heat rejection condenser   The final heat exchanger in a multiple-effect evaporator where water, usually seawater, is used to condense the water vapor produced in the last effect.

HSD   Homogenous solution diffusion.

Infiltration   The movement of water into and through a soil.

IOC   Inorganic chemical.

Ion trap   A vessel containing aluminum shavings, which exchange ionically with heavier metals.

Jet ejector   See Eductor.

Latent heat   The energy stored or released by a substance as it undergoes a physical change of state; e.g., ice melting to water or water boiling to steam.

Limiting current density   The current that causes polarization, when the current available to transfer ions in electrodialysis exceeds the number of ions available to be transferred.

LSI   Langelier saturation index

Mass transfer   The passage of a given mass of material as for example, through the membrane to the permeate side.

Mass transfer coefficient or MTC   Coefficient quantifying material passage though the membrane. MTC of water typically stated in gsfd/psi or day; MTC of solute typically stated in lb-sfd/psi or day.

Membrane element   A single membrane unit or cartridge.

Membrane pair   A combination of one anion membrane and one cation membrane in electrodialysis and electrodialysis reversal.

Membrane system   Desalination of water using a semipermeable membrane.

Micron   Unit of measure equal to 10^{-6} meter.

Multiple-effect distillation (MED)   An evaporative process consisting of several stages in series by backing of a thin film of saline water flowing over a surface of heat transfer tubes or plates capitalizing on the temperature/pressure boiling curve characteristics of the liquid.

Multistage flash distillation (MSF)   An evaporative system that uses flash distillation of a saline water solution in several stages to capitalize on the temperature/pressure boiling curve characteristic of a liquid.

NOM   Natural organic matter.

NTU   Nephelometric turbidity unit.

Nonalkaline scale   See Hard scale.

Osmosis   The spontaneous transport of water through a semipermeable membrane from a solution of low salt content to a solution of high salt content to equalize salt concentrations.

Performance ratio   A measure of the pounds of product water produced per pound of motive steam applied to the process.

Permeability   The capacity of a membrane to allow water or solutes to pass through.

Permeate   That portion of the feedwater which passes through a membrane.

Permeator   A reverse osmosis production unit consisting of the membranes and pressure vessel.

pH   A logarithmic scale describing the concentration of hydrogen ions in a solution.

Phase reversal   An electrodialysis process where the polarity on each membrane stack is reversed in a timed sequence.

Polarity   The direction of current flow in an electrical process.

Polarization point   The point at which the current density is high enough to dissociate water.

Polyamide   An organic polymer containing NH_2 functional groups.

Polyphosphates   A polymer containing PO_4 linkages.

Porosity   The proportion, usually stated as a percentage, of the total volume of material that consists of pore space or voids and is open to fluid flow.

Posttreatment   Processes that may be used on the product water, such as disinfection, demineralization, carbonation, degasification, to make the product suitable for use; neutralization, posttreatment of concentrate also may be required before disposal, such as pH adjustment, degasification.

Precipitate   A substance separated from a solution by chemical or physical change as an insoluble amorphous or crystalline solid.

Pressure vessel   A single tube designed to contain pressurized fluid and hold one or more membrane elements in series.

Pretreatment   The processes such as chlorination, clarification, flocculation, coagulation, acidification, and deaeration that may be used on feedwater to a desalting system to minimize algae growth, scaling, fouling, and corrosion.

Product water   Water that has passed through a water treatment plant, i.e., water for which all the treatment processes are completed or finished. This water is the product from the water treatment plant and is ready to be delivered to the consumers. It is also called finished water.

Recovery (R)   The amount of product water divided by the amount of feedwater, generally reported as a percentage $(Q_p/Q_f) \times 100\%$.

Rectifier   A device that converts alternating current into direct current.

Reject   See Brine, Concentrate.

Retentate   The stream existing in a membrane device, which has increased concentration of solutes and particles over the feedstream.

Residual   The amount of an item that remains, unaffected, after a chemical or physical process occurs.

Reverse osmosis (RO)   The transport of water from a solution having a high salt concentration to a low salt concentration solution through a semipermeable membrane by applying pressure in excess of the solution's osmotic pressure to the solution having a high salt concentration.

RIB   Rapid infiltration basin.

Salt rejection   The selectivity to exclude dissolved ions from passing through the membrane; measured as (100-salt passage) expressed as a percentage; 100 (1-product concentrations divided by feed concentration).

Saturation   The maximum amount of material that can be dissolved in solution without forming a precipitate; the maximum amount of energy a compound can have without physically changing states.

Scale   Any material that forms a hard solid on surfaces, usually sparingly soluble salts such as calcium sulfate, calcium carbonate, or magnesium hydroxide.

Scale inhibitor   An agent that is added to feedwater to extend the limits of saturation of scaling substances. The agent ties up and inactivates certain metal ions. Also used as a cleaning agent.

Scaling   The precipitation of inorganic salts on the feed side of the membrane, pipes, tanks, or backer condensate tubes.

SDI   See silt density index.

Semipermeable membrane   A membrane that preferentially allows the passage of specific compounds while rejecting others. For example, reverse osmosis membranes will allow water to pass, but not salt.

SHMP   Sodium hexametaphosphate.

Silt density index   A procedure that determines the concentration of colloids in feedwater derived from the rate of plugging of a 0.45-μm filter at 30 psi (107 kPa) applied pressure.

SOC   Synthetic organic chemical.

Soft scale   See alkaline scale.

Solids rejection   Percentage of mass removed from the feed, $(1 - C_p/C_f) \times 100\%$.

Solubility   A measure of the amount of a certain substance that can dissolve in a given amount of water at a given temperature.

Solute   As defined here, a dissolved substance to be removed by the membrane.

Solution   A homogeneous mixture of substances in which the molecules of the solute are uniformly distributed among the molecules of the solvent.

Solvent   The liquid medium that carries dissolved substances typically water.

Stage or bank   Pressure vessels in parallel.

TDS   Total dissolved solids.

THM   Trihalomethanes.

TOC   Total organic carbon.

Train or array   Multiple stages in series where typically the concentrate is used as feed to the subsequent stage.

Turbidity   Any undissolved materials in water, such as finely divided particles of sand or clay, that reduces the penetration of light and causes the water to appear cloudy.

Vapor compression evaporation (VCE)   A desalting process in which heated saline water is introduced as a fine spray on the outside walls of heated tubs in a chamber for which the pressure is lower than required for vaporization, causing some of the water to vaporize. The tubes are heated by the compression of the vapor collected from the chamber and subsequent condensation of distillate (product water) inside the tubes from compressor discharge. Typically the vapor is compressed by using a mechanically driven compressor or a steam jet thermocompressor. Units of one to four stages are common.

Water transport   The passage of water through a membrane. Water transport is desirable in reverse osmosis and undesirable in electrodialysis.

Index